MEDIAWARE
Selection, Operation
and
Maintenance

•

Raymond Wyman

MEDIAWARE

MEDIAWARE

Selection, Operation and
Maintenance

•

Raymond Wyman
University of Massachusetts

WM. C. BROWN COMPANY PUBLISHERS
Dubuque, Iowa

Printed in the United States of America

Preface

This book is written for the people in education who must understand, select, operate and maintain audiovisual equipment, or train teachers and students in its operation.

It is far more than an equipment operation manual. It is designed to provide audiovisual specialists, teachers and training directors with enough technical information so they can understand how the equipment developed, its reason for being, the various characteristics to consider in selecting equipment for specific educational settings, and the maintenance procedures that will make for better operation and longer life.

Some sections may appear very technical to many of the people who are involved with audiovisual equipment characteristics and capabilities for the first time. However, one of the facts of modern educational life is the dependence upon technical devices for doing many things that need to be done. Every worker is faced with learning about new tools for doing his job, and educators are no exception. This book is written with the assumption that most readers have a very limited or no technical background, so much use is made of analogies and simplified explanations. Technically trained people will want to skip over much of the background material and concentrate on the applications of it to solving educational problems.

No step-by-step picture stories on the setting up and operating of specific brands and models of equipment are included. There are several reasons for this omission. Equipment is changing rapidly as industry responds to the demands of education for better and more specific tools. The number of brands and models has expanded so much that space would not permit this approach even for the most common machines. Presently manufactured machines are supplied with much more complete labels and instructions than in the past. Edu-cators are now much more knowledgeable and sophisticated about audiovisual equipment operation, so that they can apply general principles to specific machines.

This book is intended for several purposes in the general area of audiovisual equipment selection, operation and application.

It is intended for the laboratory part of a basic college or university audiovisual or media course for teachers or teacher trainees in conjunction with lecture-demonstrations and one of the standard audiovisual texts listed in the reference section. This use would require one two-hour laboratory session each week for a semester in a room that included representative pieces of the equipment to be studied and operated.

It is intended for a complete course in educational technology for professional audiovisual specialists who must direct or supervise a media program and train teachers, aides and students in machine operation. It would have particular application in the various institutes designed to increase and improve teacher competencies in media. For these purposes the book would be used in conjunction with lecture-demonstrations on equipment characteristics and capabilities, service and instruction sheets for each category of equipment, and the operation sheets would be used for comparing various brands and models for various applications.

It would also be useful in a new course in audiovisual equipment at a junior college, community college or technical institute that prepares technical assistants to support the media program at an individual school or school system. There is much need for two-year trained personnel, particularly to support the professionals on a team teaching program. Technical media personnel are also in increasing demand at colleges, universities and educational-television centers.

It would also be suitable for an audiovisual laboratory experience as part of a general or special educational methods course at a college or university that included a course in physics or physical science as a prerequisite or where some previous audiovisual experience could be assumed.

All of these uses would require a laboratory setting that included representative pieces of the equipment to be studied and operated, materials to be used with the equipment, a file of the instruction sheets provided with each machine (duplicates can be obtained from the manufacturers), at least two twenty ampere electrical circuits and several outlets, and a technically competent instructor or supervisor. A team of two people on each machine or assignment is recommended. Only one of each category of machine will suffice if teams are rotated through the series. If the machines are individually mounted on wheeled projection stands or carts, they can be stored out of the way and the room can be used for other purposes at other times. Carrels or booths provide more privacy for the teams and permit keeping related things together, but require a more specialized room.

Since standard audiovisual machines are used in the laboratory, it would seem logical to use them for other periods of time to support the regular instructional program. This is not recommended. These machines will be tampered with regularly by inexperienced people which makes them unreliable for ordinary classroom use. A separate inventory of equipment for the laboratory is urged.

Contents

Part I

Introduction
and
Technical Background

chapter 1

Introduction

A limited amount of research and a great deal of practical experience indicate that the wide and wise use of appropriate media and equipment for using them make significant contributions to the teaching-learning process in a great variety of situations.

Several conditions must be met in order to have this wide and wise use of media actually make their potential contributions.

Unquestioned competence in the subject matter to be taught is required. No one communicates any ideas, principles or facts that are not first clear to him.

An appropriate place is required. A discussion can be conducted in the middle of a field, and a lecture can be delivered almost anywhere that is reasonably quiet, but a presentation involving a variety of audio and visual experiences requires a specialized area that has been created or adapted for the purpose. A section on room conditions for easy and effective media use is included in Chapter 5.

A library of appropriate book and nonbook materials called media or software must be available to the teacher for preview, study, consideration and selection. A library of them with a common card catalog needs to be near and convenient for use when they are wanted.

A local production center should be available so that specialized materials not in the media library can be made as required. These materials are ordinarily not as competently prepared as the commercial materials, but in the hands of the teacher who suggested or designed them they can be very effective. Extemporaneous materials are often needed to capitalize on situations that arise in class. Individual students or small groups of students should also be able to present their ideas, questions and summaries to their peers with audio and visual materials made in a local production area.

Mediaware (audiovisual equipment or hardware) must be available so that materials can be presented or reproduced when and where they are needed and with a minimum of bother and distraction. Regularly used equipment should be a permanent part of the room. It is interesting to note that chalkboards became popular only when they became permanent classroom appliances. Equipment for special presentations should be available nearby on wheeled stands or delivered by someone on request. No teacher desiring to use a film should be required to find, schedule and transport the projector, then find something to put it on, then find and set up a screen and so forth.

Effective media use also requires time in the school for selection, preview, adaptation and tryout. A book or paper can be read or checked at home, but media presentations must ordinarily be prepared in the school or a special area near where they will be used. Secondary and college teachers have traditionally had free periods during the day for preparation. One of the important advantages of team teaching in the elementary school is the time that becomes available for preparing media presentations.

TERMINOLOGY

The terms "audiovisual" or "audio-visual" equipment and materials came into general use during the thirties and reached a high level of acceptance during and immediately after World War II. There has never been any agreement whether the hyphen should be used or not. This book, for no good reason, does not use the hyphen.

Because of the limited application of motion and still pictures in a mechanical way to elementary educational tasks and goals by poorly trained practitioners, the term audiovisual education has been abandoned by many professionals in the field in favor of such terms as instructional technology, learning resources, media, instructional materials, etc. There is as yet no complete agreement on a broad term to substitute for audiovisual education.

Media has gained wide acceptance as the designation for the print and non-print materials now used in education. Media centers, media libraries, media courses and media specialists are now a regular part of the educational scene. Federal legislation and support programs have made considerable use of the term "newer media" to refer to educational materials other than books.

Ware is defined by Webster as "the sum of articles of a particular kind or class" as in hardware, woodenware, tinware, etc. The term mediaware has been coined for this book to include all of the equipment necessary to project, produce, reproduce, store and maintain the messages included in the media. Mediaware then becomes a replacement for the old designation, audiovisual equipment.

Mediaware includes such machines as still and motion picture projectors, cameras, sound recording and reproducing systems, television, transparency makers, etc. Media to go with the mediaware include slides, films, tapes, discs, etc. Media is the plural of medium. The mediaware has often been referred to as hardware and the media as software. They are together sometimes referred to as the medium and the message. This book is concerned primarily with the mediaware (equipment) necessary to expose or reproduce media (materials) which contain important messages for the senses of students under teacher-controlled conditions.

COMMUNICATION

Instructors or leaders are employed or selected because they have the desirable skills, attitudes, ideas, principles, knowledge, concepts and so forth that the students or trainees are expected to acquire. The purpose of the communication process is to make the desirable attributes common to the instructor and his students. If successful, the communication will result in desirable and observable changes in the individuals, usually referred to as learning.

In order to communicate skills and concepts it is first necessary for the instructor to prepare or construct selected experiences or cues to present or expose to the senses of his students. These experiences must be perceived or observed by the students through one or more of their senses. The students must next react or respond to the cues in terms of their past experience, knowledge and needs and construct a new framework, understanding and skill related to the teacher's goals. Feedback in the form of inspected or sampled responses to the material presented is used by the instructor to evaluate learning and to determine rate, repetition and difficulty of the next presented material. This teaching-learning cycle is diagrammed in Figure 1.1.

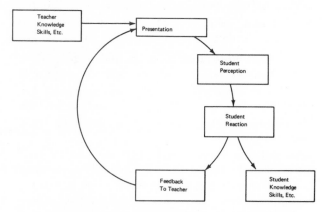

FIGURE 1.1. Teaching—Learning Cycle

The traditional and most common way to attempt to communicate material to students is through the use of written and spoken words or verbal communication. Words have several important characteristics which have made them the mainstay of classroom communication. They are so easy to produce, store, transport and reproduce. They represent the highest level of intellectual activity and have a high degree of acceptance among educated people. Words enable people, who already know what they mean, to communicate rapidly, effectively and inexpensively. Teachers are outstanding examples of experts in the field of verbal communication.

Many students, on the other hand, have a great deal of difficulty in constructing accurate and meaningful reproductions of the teacher's skills and concepts from the verbal materials presented to them in written and spoken form.

Most words give no clues to their meaning. *Chair* gives the five senses of the students no clues to what it stands for. Meaning can be taken *from* the arbitrary symbols only after meaning has been taken *to* the symbols. It seems obvious that schools should be places where students have experiences in connection with the words that represent them.

Verbalism is a term often used in audiovisual discussions to describe the use of words that are not properly understood. The humor sections of teachers' magazines are filled with examples of students' faulty, erroneous or ridiculous use of verbal language.

Teachers commonly overestimate the experiences that their students have had in the areas being taught. (Teachers are as often ignorant of the experiences that children from other cultures and regions have had.) The students who have not travelled to, thought about, touched, seen, smelled and tasted the things described in words by the teacher may seem dull, uninterested or even antagonistic. Audiovisual experiences in the classroom can often substitute effectively for real experiences.

Many words have several meanings, and the meaning intended by the teacher may not be the meaning recalled by the students. This language problem is called referent confusion. The teacher may use the words *block* or *line* with one meaning and any of the dozen other meanings may be recalled and referred to by students. Simple visuals or sound effects are very effective devices to indicate the intended referent.

Words with entirely different spellings and meanings sound exactly alike and may result in errors. The clamp used to hold materials together could just as accurately be copied in student notes as *C* or *sea*. *Eight* and *ate*, *two* and *to*, and *bale* and *bail* are other examples.

Auditory discrimination and poor articulation cause many problems with words that sound nearly alike. Many cartoons are constructed on the theme of confusion between words with entirely different meanings that sound almost alike. Accents in different parts of the country cause similar confusion.

Idioms in which a meaning unrelated to or remotely related to the actual meaning of the words cause much confusion, often to the merriment of those who understand and the discomfort of those who do not understand. Every culture and subculture has its long list of idioms which are not understood by outsiders. Neither teachers nor their students should be "outsiders" in using an idiom.

The constant, accurate and meaningful use of verbal communication is the goal desired in most educational situations. In order to arrive at it, multisensory and particularly audio and/or visual experiences need to be a regular part of the learning process. At the end of a unit of work, verbal materials should be far more meaningful than at the beginning.

ADVANTAGES OF AUDIOVISUAL PRESENTATIONS

The major advantage of audiovisual presentations lies in their ability to aid accurate and meaningful communication when used with or instead of words. There is less verbalism or parroting of words when multisensory experiences are provided in the classroom under the guidance of an instructor.

Students remember longer the presentations they have seen and heard in a realistic manner rather than merely from listening and reading. More permanent learning is an important contribution of educational materials to education. The usual reasons for forgetting are lack of understanding, lack of application to relevant problems and no feeling of importance. Audio and visual experiences can help make complex concepts clear, show the applications or usefulness of them and provide an indication of their importance.

Students show a greater interest in the audio and visual experiences that can be provided in classrooms. They are confronted with a great variety of high-quality media everywhere else in their lives, and similar variety and quality of media are needed to interest them in schoolwork. Comments of students entering a classroom obviously set up for audiovisual use reflect their interest in having material communicated in this fashion.

Emotional impact or involvement is necessary to change or establish many important ideas or attitudes. It is easy to change a person's skills or verbal answers, and very difficult to change his behavior or feelings when out of class. Pictures taken of children with infrared (invisible) light during well-projected films often dramatize the high degree of personal involvement that can result. Most adults will readily admit how moved they have been by certain films. Various media can be used to establish a mood or disposition to do something different.

Greater interest and emotional involvement result in increased ability to enter into discussion and willingness to accept responsibility for assignments and further involvement.

Audiovisual materials can often be used to teach something in less time than by strictly verbal presentations. Numerous lessons have been converted to television, which is essentially a complete audiovisual presentation. The typical forty-five or fifty-minute lesson usually requires about thirty minutes for effective presentation over television. Some of the time is saved by more effective communication techniques, and some by elimination of nonessential activities. An alternate claim can be made for teaching more material in the same time.

Although audiovisual instruction saves time in presentation, it requires extra or specialized preparation time. Preparation for a lecture can be done in many places, including the library and livingroom. Preparation for an audiovisual presentation must be done where equipment and materials are available, which is usually the instructional materials area or classroom. It can be expected that some teachers will develop home laboratories for previewing, preparing and adapting materials for class presentations.

MATERIALS OR SOFTWARE

Audiovisual equipment ordinarily requires materials such as films, filmstrips, slides and recordings in order to communicate important messages to specific groups of students. These materials are not covered in this book. The equipment is general in that it may be applied to a great variety of teaching situations. The materials are more specific in that they are designed to teach a restricted group restricted subject matter from a restricted point of view in a restricted way. There are still many possible and desirable variations within the restrictions. Many catalogs from producers and distributors list and describe available materials to go with the machines. Several directories are available that list most of the materials according to type (film, filmstrips, etc.) or according to subject and grade level. Most journals in the subject disciplines now include advertisements of current materials and authored articles or departments in these often evaluate materials. (These should all be available at the audiovisual center or laboratory.)

In a methods course a follow-up assignment to each piece of equipment that requires materials might be to prepare an annotated list of most needed materials for a specific unit of work to be taught under specific conditions. One of these materials might be previewed and a complete lesson plan prepared. A microlesson involving equipment and material might be prepared and taught.

EFFECTIVE USE OF AUDIOVISUAL MATERIALS

Everyone has seen numerous examples of irrelevant or ineffective use of audiovisual materials. The most common case is substituting an available prepared audiovisual presentation for an unprepared but needed lesson. Many lessons have been prepared on tape, film or disc that will interest a group of students even though they have no, or very limited, bearing on what was supposed to be taught at that time. The only legitimate use for audiovisual materials is to teach better what was supposed to be taught anyway. Ineffective use often involves poor preparation, poor presentation, poor follow-up activities, or a combination of all three.

Good audiovisual education requires appropriate material, equipment, setting, timing and lesson plan.

Audiovisual materials have been made in tremendous numbers, and it is important to select the most appropriate material from that available for the job that needs to be done at reasonable cost. Every audiovisual center has catalogs, indices, lists and descriptions of its materials that can be consulted. Many centers distribute descriptive materials that can be consulted without visiting the center. Producer and distributor catalogs can be studied for purchase, rental or loan of distant materials. Most professional journals now carry descriptions of new curricular materials.

Some criteria for selecting materials may help match them to specific needs.

The subject matter should agree with the curriculum. It cannot be stressed too much that the curriculum should determine content and materials rather than allowing available materials to determine curriculum. Many publishers of printed materials prepare audiovisual materials that are correlated with and an enrichment to them.

The grade or maturity level should be correct for the audience. It is primarily the vocabulary used that determines grade level. The same picture can be used with a wide grade range if it is described in appropriate language. Sometimes it might be desirable to use a sound-picture combination with the teacher's narration rather than the one provided.

The material should be accurate and authoritative. This can most easily be determined by checking the collaborator or author, who should be an authority in the field. The date of production should also be checked in rapidly changing fields. If small errors or omissions are noted, these can be pointed out before, after or during use.

The material should be structured for logical, easy and effective teaching if the material is a film, filmstrip or recording in which the parts cannot easily be rearranged. Audiocards, slides and transparencies can of course be structured or sequenced as desired.

The important ideas should stand out and be clear to the intended audience. Special note should be made of distracting elements. Key words and diagrams should be included whenever needed.

Technical excellence should be required so that all can see and hear the details needed.

The material should be attractive. Good photography, graphics and sound will enhance any production. Although color seldom adds to learning in controlled experiments, there is a general preference for it even though it generally doubles the cost of the material.

The material should promote additional activity on the part of the students rather than merely provide them with facts. Some materials are called "open end" because they present and develop a situation and then end with a question or problem to be answered or solved by the individual or group. Good audiovisual materials often promote library study, experiments, interviews and so forth.

Proper selection and use of equipment to expose the selected materials is the major intent of this book. The materials can in most cases appear no better than the equipment permits.

The setting for use of audiovisual materials and equipment is normally the classroom, an individual study carrel or an auditorium. The lighting, acoustics, electrical circuits, conduit and so forth, in these areas that need to be considered for easy and effective use are considered in Chapter 5. These details need to be provided for positively and never left to chance.

Timing refers to the optimum point in the lesson when audiovisual experiences are needed. All the preceding factors should be controlled so that the audio and visual impact needed can be delivered to the students at the best moment for them. Few media-supported programs are as yet able to provide this.

A four-step lesson plan has proved very effective in the use of media.

The first step is called teacher preparation. The teacher previews the selected material, consults the teaching guide, lists words that need emphasis or explanation and determines whether it is to be used as a whole or only in part. The proper time for presentation is determined. Supplementary materials may need to be assembled or consulted. The equipment and room arrangements need to be checked.

The second step is student preparation. The teacher introduces the material, relates it to past and future activities and proposes questions that can be answered by the medium. Key or difficult words are often put on the chalkboard or overhead projector. A diagram or map used in the medium may be sketched for later completion. Every student should have an understood responsibility as he sees or hears the presentation.

The third step is presentation in which the material can be studied easily and without distraction. It should be the center of attention. It should start at the beginning and not in the middle of the first idea. Sound and images should be as nearly perfect as the state of the art allows.

The fourth and final step is called follow-up. Just as soon as the machine is stopped, and before rewinding, packing and room rearrangement, the students should be involved in discussion, writing, thinking, searching and questioning. The material is not an end in itself, but a springboard to all kinds of valuable activties that may require considerable time and effort.

After the material has been used, a brief evaluation for future reference should be made and filed.

APPLICATION

It will be noted that the application of audiovisual equipment to specific educational purposes is conspicuously lacking in this book. This is intentional. There are so many ways to use this equipment with so many materials in so many subject and grade situations in so many settings and with various numbers of students that application becomes a very specialized professional job. It is assumed that people in a wide variety of educational situations who know their students, their subject matter, available materials and the general methods for teaching also need to know enough about the characteristics of audiovisual equipment to use it easily and effectively. A follow-up assignment to each piece of equipment studied in this book might be to apply it to specific needs for a particular grade, subject and setting. After a group in a subject area methods course has become familiar with certain pieces of equipment, the various applications might be discussed in detail.

chapter **2**

Technical Background

ELECTRICITY

Most audiovisual machines are powered by electricity, although some special purpose devices may be hand or spring operated. An elementary knowledge of electricity and electrical circuits is necessary to select, operate and maintain them.

The electron is one of the basic building blocks of matter. All atoms of all elements consist of tiny electrons spinning around a nucleus with the number generally determining which element it is. The electron is a negatively charged particle, and it is ordinarily balanced by positive charges in the nucleus. When charges are not balanced, there is a possibility that electrons may move from one atom to another. The familiar spark from rubbing a cat or walking on a carpeted floor in dry weather is evidence of electrons in motion, which is called electricity.

In order to move electrons in appreciable quantities it is necessary to provide energy in some form and to have a circuit which will permit the easy flow of electrons. Electrical energy may be produced from chemical, mechanical, light and heat energy, and all but the last method are regularly used in audiovisual equipment (see Figure 2.2).

Batteries make use of one or more cells which use chemical energy to produce a flow of electrons. The most common cell is the carbon-zinc type which is used in flashlights and countless other devices. The cells may be combined in various ways to produce batteries with various electrical characteristics. All cells consist of two unlike materials such as carbon and zinc and a conducting fluid called an electrolyte which may be absorbed and sealed as in a dry battery, or free to move around as in a wet battery.

The carbon-zinc battery is most common because of its simplicity, low cost and reliable performance. It is diagramed in Figure 2.1. One electrical termi-

nal is ordinarily the capped end of the central carbon rod and the other is the zinc case which may be enclosed and sealed in steel. The most common cells are designated with letters such as AA for penlights, C for small flashlights and D for large flashlights.

Two other dry cells designated as mercury and alkaline are gaining rapidly in popularity because of greater capacity per unit of volume and weight,

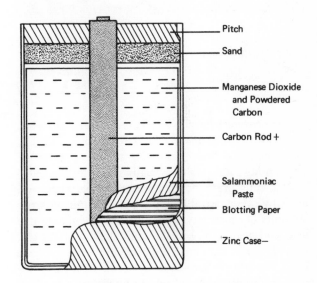

FIGURE 2.1. Carbon—Zinc Cell

although at higher unit cost. When small size, low weight or long life are particularly important, then these cells should be considered. Alkaline cells can ordinarily be interchanged with carbon-zinc, but mercury cells have reversed polarity and other differences.

Dry cells may be recharged by applying electricity properly to their terminals with special charging equipment. The procedure tends to reverse the chemical action that produced electricity.

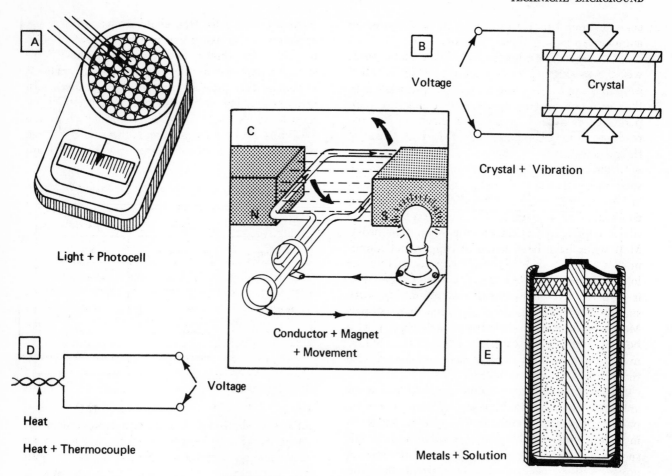

FIGURE 2.2. Ways To Produce Electricity

However, recharging dry cells is not very efficient, it cannot be done more than a few times, and it is a bother when new ones are available at low cost. Few audiovisual people actually recharge dry cells.

Wet cells are used in automobile batteries and in many other portable applications requiring large amounts of electricity. Wet batteries commonly use lead and acid or nickel-cadmium-alkaline. Both can provide large amounts of electricity, and they can be efficiently recharged many times with little deterioration. Both liquid electrolytes are very corrosive and precautions must be taken. Most automobile batteries contain lead and an acid which can be inspected by removing one of the top caps. Most high-power portable electric equipment such as television and cordless drills now make use of newly developed sealed nickel-cadmium batteries with high performance and easy recharging characteristics.

Electricity is produced whenever a piece of wire moves in a magnetic field or a magnetic field changes in the presence of a wire. This is the generator principle that is used to produce practically all the power used in homes and industry. Turbines rotated by water or steam power move many large wires in magnetic fields (generators) to produce electrical power. The automobile generator operates from the engine to charge the battery and supply power to various electrical devices. Various microphones, phonograph cartridges and tape recorder heads produce very small amounts of electricity according to this principle for amplification and reproduction as sound or pictures.

Electricity is also produced by mechanically vibrating various crystals that have metallic foil cemented to their surfaces. This is called piezo electricity and the principle is commonly used in inexpensive microphones and phonograph pickups with the vibration coming from sound waves or modulated grooves on the record.

When light energy strikes certain chemicals specially arranged in what are called photocells, electricity in small amounts is produced. This conversion of light energy into electrical energy is used in photographic exposure meters, television cameras and to obtain the sound from a sound track in most common sound movie projectors. It is also used in automatic focusing slide projectors.

Heat applied to a junction of unlike metal wires will also produce electricity and the device is called a thermocouple. Its most frequent application is in thermometers which may be used to measure temperature on a slide or inside a machine. Remote reading room temperature devices make use of thermocouples.

Electrons or electricity flows much better through some materials than others. All the common metals are good conductors, and most nonmetals are poor conductors or insulators. The best conductors are silver, copper, gold and aluminum, in that order. Most wires have been made of copper, but aluminum is gaining, particularly for large capacity over long distances. With similar diameters, copper wire is a better conductor than aluminum, but with similar weights, aluminum is better than copper. Most complex circuits are made with copper wire because it is much more easily soldered than aluminum. Copper wire is, in addition, often coated with tin to make it easier to solder.

Some of the most effective electrical insulators are glass, porcelain, bakelite, rubber, beeswax, paraffin, dry wood and cotton. Pure water is an insulator, but when contaminated with materials such as inorganic salts and acids, it becomes a good conductor. Dry air is an excellent insulator. Most wires used in audiovisual equipment are now insulated with one of the new plastics such as polyethylene or vinyl.

The conductivity of a wire is improved by increasing its diameter, decreasing its length and decreasing its temperature. Wires are often made of many small strands in order to increase flexibility. Insulation is usually improved by making it thicker.

Electricity is measured in units which are most easily understood by comparing it to water as in Table 1.

It takes pressure to force water through a pipe, and it is measured in pounds per square inch. Ordinary household tap water is forced out by a pressure of about 100 pounds per square inch. A lower pressure will force less water through the pipes and a higher pressure will force more. If the pressure falls to zero then no flow will occur.

It takes electrical pressure called voltage to force electrons through a wire or circuit. A zinc-carbon dry cell provides one and a half volts no matter whether it is large or small. Several cells may be connected in series to obtain higher voltage. Most automobiles have 12-volt systems with batteries

made up of six cells. Household lamps and portable appliances such as audiovisual equipment operate on 120 volts. Stoves and other permanently installed equipment usually operate on 240 volts. It is voltage that provides the shock associated with electricity. A low voltage such as that provided by most batteries cannot even be felt. Many years ago 120 volts was chosen as the highest safe voltage that might be used in homes for portable ap-

TABLE 1. Water—Electricity Analogy

	Water	Electricity
Pressure	lbs. per sq.in.	volts V
Flow	gal. per sec.	amperes A
Resistance	friction	ohms R
Work	pressure x flow	watts W
Power	horsepower	kilowatt hours KWH

pliances. Although 120 volts will provide a very unpleasant shock, it is seldom dangerous unless the person happens to be wet, in which case it may be fatal. Extreme precautions should be taken with 120-volt equipment when any person might be wet. Many European countries use 220-volt househo d systems and American equipment cannot be used unless specially adapted. Some countries use a 100-volt system and much 120-volt equipment will operate, but at reduced efficiency. Voltage is measured with a device called a voltmeter and several models will be found in any science laboratory. Equipment nameplates, Figure 2.3, often indicate a range such as 105 to 125 volts over which satisfactory operation can be assured.

All batteries and the original home electrical systems operate on direct current (DC), which means that the electrons always flow in one direction. Direct current has the advantages of simplicity,

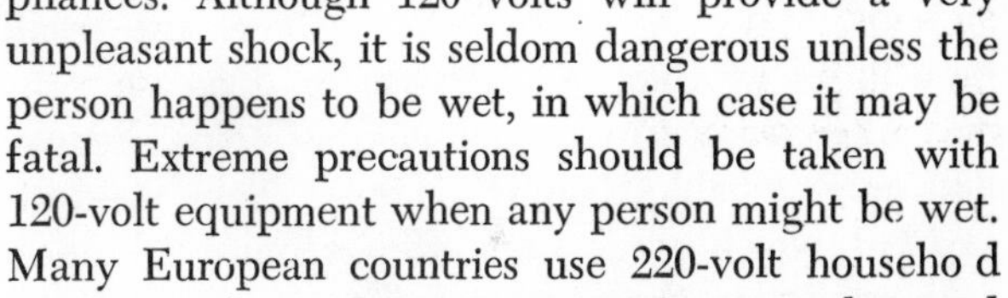

ACME SUPER

Model 1000   Serial No. 2658

105-125 V. 60 ∿   1000 W. with 750 W. Lamp

(UL) and CSA Approved. Use DDB or DGS Lamp

FIGURE 2.3. Equipment Identification Nameplate

silence (no hum) and direct connection to batteries.

Almost all household and institutional electricity in use today is alternating current (AC) which flows first in one direction and then in the other. This is diagramed in Table 2. It is called sixty-cycle alternating current because it completes sixty excursions from zero volts to about 168 volts with the electrons flowing in one direction, back to zero, and then to 168 volts with the current flowing in the opposite direction. This alternating current will do the same work done by 120-volts direct current. It is often called 120-volts root-mean-square (RMS). Cycles per second is being replaced in much of the literature with Hertz (a famous scientist), so cps = Hz. Sixty Hz is also abbreviated 60 ~. Separate meters and equipment must usually be selected for AC or DC. However, incandescent lamps work equally well on either.

Alternating current can be stepped up in voltage with transformers for economical transmission over long lines and then stepped down with other transformers for safe and convenient use. Direct current cannot be transformed and it is used only in local systems. Many individual AV machines use small built-in transformers to increase or decrease voltage. Alternating current motors are simpler, cheaper and more reliable than DC motors. They need no brushes which wear out, often require attention and cause sparking which may cause radio or audio interference. Most AV equipment that includes a motor or an amplifier is designed for AC only. Such equipment not only will not run on DC, it will probably blow a fuse or be ruined if plugged

into DC. Ordinary projection lamps work equally well on AC or DC.

It must be emphasized that DC current is still available in the center of some old cities, or at camps or on boats, and that the wall plugs may be the usual kind, and no labels may be provided. It can ruin the usual equipment designed for AC-only operation.

If there is any doubt whether the electricity is AC or DC, a custodian should be asked. Special pocket testers are available for determining AC or DC. Another way to check for DC current is to screw out slowly a 100-watt incandescent lamp that is burning. On AC it will "go out like a light"; on DC it will slowly dim with an audible arc or sparking inside the socket.

The dangers of 120-volt electricity can be minimized by the use of a third wire from the case of the appliance direct to the plumbing (ground) of the building. Figure 2.4 shows the three-contact

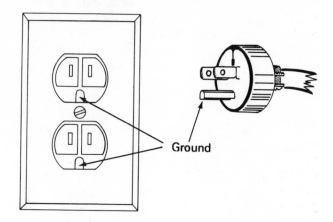

FIGURE 2.4. Three Contact Electric System

TABLE 2. 120 Volt 60 Hz Alternating Current

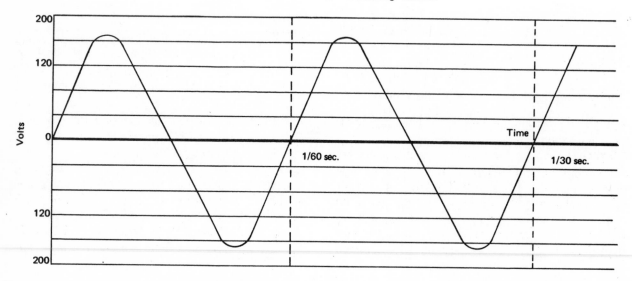

grounded duplex outlet and the matching cord and plug which are becoming standard for audiovisual operation. All specifications for new equipment and buildings should require this system. Many older buildings need to have the new grounding duplex outlets installed.

The common two blade power plug on older electrical equipment can be inserted in the wall socket in either of two ways. Hum and shock can often be decreased or eliminated by reversing the plug in the socket. In addition to getting the two blade plug inserted properly to reduce hum and shock, it is sometimes desirable to attach an external ground wire directly from the equipment frame to a water pipe. The power cord should be disconnected while this connection is made.

Adaptors with a short wire are supplied or available to use with three-contact equipment and regular two-contact outlets. The short wire on the adaptor (usually green) should be attached under the screw that holds the cover on the duplex outlet to ground the equipment. The regular two-contact plugs will fit the three-contact sockets without any adaptor.

The flow of water through a pipe (Table 1) is measured in gallons per second. The flow of electrons through a wire is measured in amperes (A) with a device called an ammeter. Ammeters come in AC or DC models. Household fuses or circuit breakers are normally rated at 15 amperes which means that they will melt, "blow" or trip at greater current and interrupt the circuit. Automobile fuses may be rated from a fraction of an ampere to 30 or more amperes at low voltage. School duplex outlet circuits are normally fused for 20 amperes, or 30 amperes in auditorium projection booths. Fuses must be replaced and circuit breakers can be reset. Some automobiles have an ammeter on the instrument panel to indicate how much current is flowing into or out of the battery. Much electronic equipment uses small currents measured in milliamperes (MA). 1000 MA = 1A.

Batteries or cells may be combined in parallel (plus terminal to plus or positive terminal and negative or minus terminal to negative) in order to increase the available current. Positive terminals are often colored red.

Switches used to interrupt electrical circuits are commonly rated and stamped with the permissable amperes and voltage they will safely and repeatedly handle. A relay is an electrically operated switch. This device permits the use of small wires and low voltage to turn on or off a distant circuit involving much voltage or current. The common home thermostat is connected to a relay at the furnace or heater.

The flow of water through a pipe is held back by friction. The analogous situation occurs in wires and the friction is called resistance (R) which is measured in units called ohms (abbreviated R or Ω). Ohmmeters are commonly used to measure the resistance of a wire or circuit. The resistance of a wire normally results in a voltage drop according to the equation $V = AR$. Ten amps through an extension cord with one ohm resistance would drop the voltage by 10 volts.

Wires for household and audiovisual circuits have an American Wire Gauge number which is inversely related to diameter and current carrying ability. Number 18 wire is .04 inches in diameter, 6.5 ohms per 1000 feet and is usually used for portable lamps, radios, television sets, tape recorders and other devices using up to about 500 watts. A common and inexpensive version is called zip cord. Number 16 wire is .05 inches in diameter, 4 ohms per 1000 feet and is usually used for flatirons (asbestos covered) and audiovisual equipment such as overhead projectors and sound movie projectors using up to about 1000 watts. Most audiovisual extension cords should have number 16 wire. Number 14 wire is used for most permanent household circuits and number 12 is most used for permanent school circuits.

Very large and very small values of electrical quantities are often encountered and prefixes are used to denote them. The following are most common:

mega = 1,000,000
kilo = 1,000
deci = 1/10
centi = 1/100
milli = 1/1,000
micro = 1/1,000,000

Common electrical quantities of interest to audiovisual engineers can be measured with a single instrument called a volt-ohm-milliammeter shown in Figure 2.5. It will measure AC or DC volts (AC/DC) but only DC ohms or amperes. In using such an instrument it is always best to start on a high range and work downward in order to avoid the risk of burning out the meter.

Moving water will do work and so will moving electricity. The work that water will do is determined by the pressure times the flow, and with electricity by the volts times the amperes which equals watts (W) or $W = VA$. The most familiar use of watts is in describing household lamps. The

FIGURE 2.5. Volt—Ohm—Millimeter

Courtesy of Triplett Electrical Instrument Co.

term is used by audiovisual personnel in describing projection lamps, the output of amplifiers and the total demand on the electrical system, usually kilowatts.

Electrical specifications often give only two of the three quantities, but the third can easily be determined from the equation. A 1000-watt projector on 120 volts would require 1000 divided by 120, or 8.3 amperes. A 15-ampere 120-volt circuit would handle a 15 times 120, or 1800-watt load.

Power involves work multiplied by time, and electricity is purchased by the kilowatt hour (KWH), which means 1000 watts for one hour.

ELECTRONICS

Electronics is the specialized branch of electricity in which the flow of electrons in a circuit is controlled by taking advantage of the characteristics of individual electrons in a vacuum ('vacuum tubes) or in certain semiconducting solids such as

germanium or silicon (solid state electronics or transistors).

The flow of current (electrons) in an ordinary electrical circuit can be controlled with a switch so that small amounts of energy applied to the switch will control large amounts of electricity flowing through a motor, lamp and the like. However, the amount of energy needed to operate the switch is relatively large, control is ordinarily limited to on and off, and the frequency of switching is limited to a few cycles per minute.

The flow of current in an electronic circuit can be controlled in almost any complex fashion desired so that power from a battery or outlet can be made to perform almost any task. The amount of energy needed for control is so small that sound waves, vibrations, magnetic fields and light can be used to determine the control. In addition to simple on and off functions, any percentage of flow can be designed into a system. The frequency of controlling the flow of electrons can easily go from a few cycles up to millions of cycles per second. The frequency, amplitude and shape of complex alternating currents can be changed for desired purposes.

Electronic circuits enable microphones, phonograph pickups, light sensitive cells and magnetic heads to control electrical power supplied by batteries or outlets in order to operate loudspeakers, make recordings, recreate pictures and the like.

The traditional method of controlling electricity is to convert it into a flow of electrons inside a vacuum tube. Figure 2.6 shows a two element tube often called a diode. The heated filament or heater plus a cathode boils off electrons in a near-perfect vacuum to form a cloud in the immediate vicinity. If a plate or anode is located nearby and it is made positive with respect to the cathode by connecting it to the positive terminal of a battery then electrons will be attracted and there will be a flow of

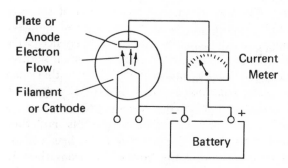

FIGURE 2.6. Electron Flow Through Diode Tube

electrons from the negative cathode to the positive anode. If alternating current from an ordinary outlet is applied to the plate, electrons will flow only during the times when the plate is positive and no current will flow during the times when the plate is negative. Diodes are regularly used in electronic equipment to convert or rectify alternating current to pulsating direct current which can then be smoothed out or averaged to approximate the electricity from a battery. This is shown in Figure 2.7. Since only one half of the alternating current is rectified, this single diode system is

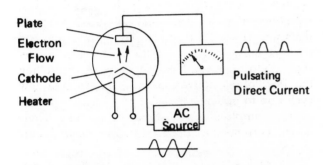

FIGURE 2.7. Rectification By Diode Tube

called a half-wave rectifier. Two diodes are often used together to form full-wave rectifiers.

Solid state or semiconductor diodes are now replacing most vacuum tube diodes in electronic equipment. Figure 2.8 shows the construction of a low power example. The solid state diode requires no vacuum, no filament or heater power, does not get hot, is much smaller, and can be made very rugged and reliable.

In order to control the flow of electrons between the cathode and anode of a vacuum tube, a spiral or screen of wire called a grid is placed between the two to form a triode tube as shown in Figure 2.9. Since the grid is mostly open space, electrons can freely flow through it to the plate if the grid is not charged. If the grid has a small negative charge on it, the electrons will be repelled back toward the cathode. If the grid has a small positive charge on it, electrons will flow to it rather than the plate, and this condition is normally avoided. In operation the grid is usually provided with a small fixed negative charge called bias, and then small changes in voltage around the bias voltage are used to control the flow of electrons through the tube. Several grids may be used in one tube

to form a tetrode, pentode and the like for special purposes.

The triode or multigrid tube is often called an amplifier because small control voltages from a microphone and phonograph pickup, for instance, can be applied to the control grid and used to regulate a substantial flow of electricity from a

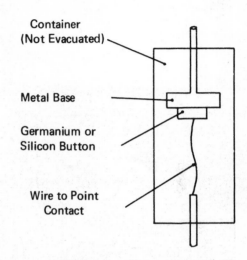

FIGURE 2.8. Solid State Diode

battery or rectified and smoothed power from an alternating current outlet. A very complex electrical wave form corresponding to the sound waves striking a microphone can be applied to the grid of a vacuum tube and an almost exact replica of it, but much increased in power, can be supplied by the plate circuit to a loudspeaker or recorder. In practice, several tubes are usually used in order to obtain the amplification or gain and the power output needed.

Solid state amplifiers and control devices make use of semiconductor diodes plus a third element, or "sandwich" which makes a transistor. The con-

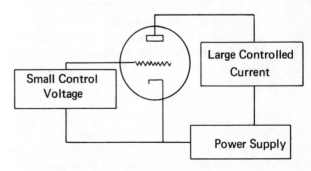

FIGURE 2.9. Triode Tube Circuit

struction and circuit are shown in Figure 2.10. Transistors are made in hundreds of configurations for various purposes. They enable a microscopic signal from something such as a microphone to control large amounts of power from a battery or rectified alternating current source to operate loudspeakers or other loads. They are usually used in multiples in order to obtain the control and power needed.

Transistors are rapidly taking the place of tubes in audiovisual equipment due to their small size, low weight, high efficiency, lack of heat and instant starting characteristics. They also operate on

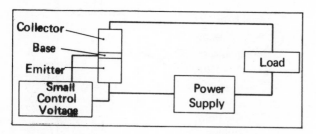

FIGURE 2.10. Transistor Circuit

much lower voltages than tubes. On the other hand, tubes can provide as much fidelity and power as solid state equipment and many applications do not require the preceding characteristics. Transistor equipment may be much more difficult and expensive to repair in the usual service shop.

Tubes, diodes and transistors are used in circuits with other components mounted on a metal chassis that often looks like an inverted cake pan, or on a bakelite or other insulating circuit board that contains etched copper bands instead of, or in addition to, wires. The most common circuit elements are resistors that impede the flow of current, condensers that permit the flow of alternating current and prevent the flow of direct current, coils that permit the flow of direct current and impede the flow of alternating current and transformers that step up or step down alternating voltages, currents and impedances. Some elements are fixed in value, and some such as volume and tone controls can be varied by the operator. Combinations of elements can be used to separate and selectively amplify certain signals or frequencies. A typical amplifier is shown in Figure 2.11.

Electronic circuits are built into record players, public address systems, tape recorders, sound motion picture projectors, television systems and even some automatic slide projectors. The purpose of this book is to supply enough information for un-

derstanding, selection, operation and maintenance. The design, adaptation and repair of electronic equipment should be left to qualified technical personnel. Dangerous voltages are used in many electronic circuits.

SOUND

Sound is that form of energy which is used by the human ear to provide the sensation of hearing. The hearing mechanism is able to detect extremely minute amounts of sound energy and distinguish among countless variations in its form.

Sound is produced by vibrating strings (violin), vibrating columns of air (organ), vibrating solids (xylophone) and vibrating diaphragms (drum). It requires some other form of energy, usually mechanical, to initiate and sustain the vibration which is converted into sound.

Sound ordinarily travels through the air from source to destination. At ordinary atmospheric pressure and 32 degrees Fahrenheit it travels at 1087 feet per second or 740 miles per hour. This seems to be a very high speed, but since light travels much faster, the sight and sound of even a moderately distant event soon get out of synchronism. Sound also travels through solids and liquids and at much higher speeds than in air. Sound will not travel through a vacuum. Several good sound absorbers or construction techniques have been found that will effectively reduce the transmission of sound.

The eye cannot ordinarily see sound waves, but an oscilloscope connected to a microphone and amplifier will display a representation of sound waves as in Figure 2.12. The pressure at any given point rapidly fluctuates up and down from normal with the progression of time. Any particular area of compression or rarefaction (lower density) proceeds outward in all directions at 1087 feet per second.

Sound waves differ in three important characteristics which can be easily identified by conventional diagrams as in Figure 2.13. In each case, one characteristic is changed while the other two remain constant.

The wave form can change in amplitude which means greater alternating compression and rarefaction of the air. This change is normally interpreted by the ear as loudness or intensity. Audiovisual equipment usually produces sound from electricity, and so the output of amplifiers is often given in watts. Since sound reproducers such as loudspeakers or earphones differ widely in their

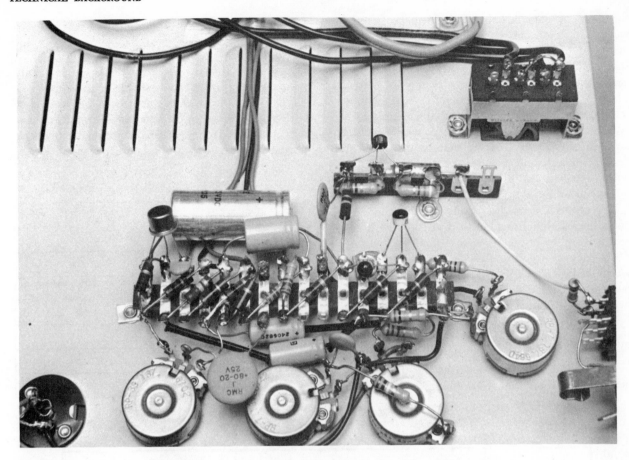

FIGURE 2.11. Typical Transistor Amplifier (View from underside)

Courtesy of Audiotronics Corp.

ability to convert electricity into sound, and since loudness decreases rapidly with distance another system of measuring loudness is needed.

It is customary to express sound levels in relative intensities with a unit called the decibel, or dB. A normal individual can just distinguish the differ-

ence in intensity between two sounds one decibel apart under ideal conditions. The threshold of hearing for normal individuals is usually considered zero decibels at 1000 cycles per second and the

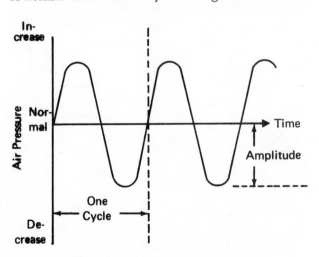

FIGURE 2.12. Visualized Sound Wave

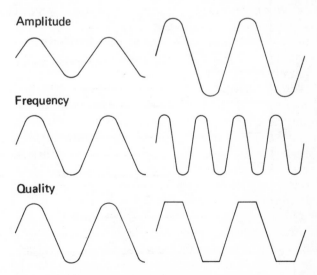

FIGURE 2.13. Sound Characteristics

threshold of pain is about 120 decibels. In other words, a person can barely distinguish 120 increments of sound intensity from faintest to loudest. It should be noted that zero decibels does not necessarily mean that no wave energy is present. The dB levels and relative power requirements of some common sounds are indicated in Table 3.

Sound intensities are measured with precision devices called sound level meters that indicate decibels directly on a meter. They are often used with filters so that the intensities of various frequency bands can be determined.

It requires energy to produce sound, but it is not a linear relationship. Instead, it is a logarithmic relationship. In order to increase loudness by three dB, it is necessary to double the power. To increase loudness by six dB would require four times the power. The relationship between sound levels in decibels and power ratios is shown in Table 4.

The decibel is a logarithmic power ratio defined by the formula $dB = 10 \log \frac{P^2}{P^1}$. In sound level measurements P^1 is considered as the threshold of hearing and P^2 is the observed power. If the ob-

served power is ten times the threshold power, then $dB = 10 \log \frac{10}{1} = 10 \log 10 = 10$, since the logarithm of 10 is 1. The logarithms of a few common numbers are given in Table 5.

The wave form can change in the frequency with which the compressions and rarefactions are repeated. The frequency has traditionally been reported in cycles per second, or cps, but a newer

TABLE 4. Decibels and Power Ratios

DECIBELS	POWER RATIOS
40	10,000
30	1,000
20	100
10	10
6	4
3	2
0	1

TABLE 3. Sound Levels

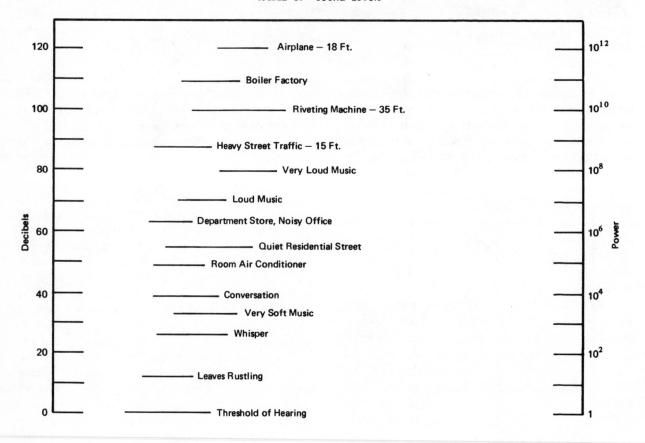

TABLE 5. Logarithms of Common Numbers

NUMBER	1	2	3	4	5	6	7	8	9	10	100	1,000	10,000
LOGARITHM	0	.30	.48	.60	.70	.78	.85	.90	.95	1.0	2.0	3.0	4.0

TABLE 6. Frequency Range of Common Sounds

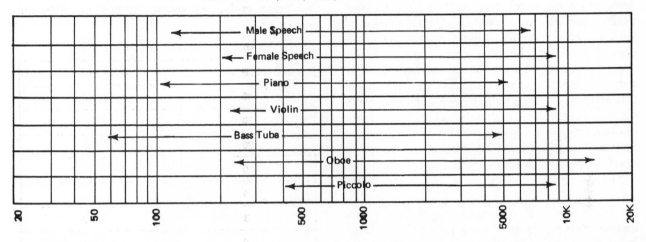

nomenclature equates cycles per second to Hertz or Hz, named after a famous physicist.

The frequency of a sound wave is the number of complete waves passing one spot in one second. The human ear can ordinarily detect as sound only those frequencies between about sixteen and sixteen thousand cycles per second or Hz. The low frequencies are called bass, and the high frequencies are called treble, and tone controls are often labeled with these terms. Neither the very low nor the very high frequencies are needed for intelligibility nor the usual enjoyment of sound, and individuals differ widely in their abilities to hear the extremes. The effect of frequency on the ear is called pitch. The frequency associated with various sound waves is shown in Table 6.

The human ear is not equally sensitive to sounds at different frequencies. It is most sensitive to sounds at about 3500 Hz and about 10 dB less sensitive at 10,000 Hz and about 30 dB less sensitive at 100 Hz.

Sound waves also differ in the shape of the wave which results in what is called quality. Exactly the same intensity and frequency of sound could be produced by a buzz saw, violin string or a horn, and the ear would easily distinguish which was which. The peculiar sounds would each have a different shaped wave form with identical amplitudes and frequencies. The different shapes are due to the presence of lower intensity but higher frequency components at multiples of the basic frequency. The higher frequencies at two, three and four times the fundamental frequency are called harmonics. Most common sounds are very complex and include harmonics throughout the entire audible spectrum.

Noise is obviously one kind of sound, but in practice it is usually used to designate irregular, confused or discordant sounds. Musical sounds are regularly recurring and pleasing sounds. It might be added that one person's music may be another person's noise.

ACOUSTICS

Sounds behave in well-known ways that affect their production, transmission and use.

Sounds ordinarily radiate from a source in all directions and the intensity decreases as the square of the distance. When one moves twice as far from a sound source in open space (not a room) the intensity drops to one-fourth, or down about 6 decibels. This means that it is very difficult to transmit sounds over more than moderate distances unless very loud sounds are produced. The intense sound of thunder is seldom heard more than a mile away, and the jet airplane that produces 120 dB (pain) at 50 feet is not audible when it is at normal altitude. Many low-power sound sources are far more economical and effective than one high-power source when large areas must be covered.

Sound bounces or reflects from most surfaces and the result is called an echo. If sound travels 500 feet to a surface and then returns 500 feet it will be delayed about one second, and of course, considerably decreased in volume. Surfaces as close as 60 feet will produce annoying echos when one is trying to understand a speech. Echos are a particular problem with outdoor public address at events such as commencement ceremonies or religious services. It is almost impossible to find a location among buildings where high-powered public address will not produce annoying echos. No satisfactory sound-absorbing material is practical for covering large offending surfaces. Some thick shrubbery is effective with low buildings.

When sound is reflected from several surfaces, usually in a large room or auditorium, the multiple echos are called reverberation and the resulting sound may be very difficult to understand. The trouble is increased as the room dimensions become greater and as large flat reflective surfaces are used. Reverberation is less troublesome when rooms are small, odd shaped, and when walls, ceiling or floor are covered with sound-absorbent materials.

The reverberation time for a room is the time it takes in seconds for a sound to diminish to one millionth of its original intensity, which is equal to 60 decibels. Classrooms with reverberation times above one second sound too "live" and those below .5 seconds sound too "dead." A rough idea of reverberation time can be obtained by listening to the decay of sound from a sharp hand clap. Too short a reverberation time is as bad as one that is too long. Radio and recording studios normally have a lower reverberation time than classrooms so that studio reverberation will not be added to reproduction room reverberation with an unpleasant total. An anechoic room is one with such extensive sound absorbency that it is considered "dead." Auditoriums have somewhat longer reverberation times than classrooms, and this is desirable.

Sound is absorbed by most soft materials added to a room and by specially treated or perforated hard materials such as ceiling tile. Each material has an absorption factor measured in sabins. Most materials are far less effective in absorbing low frequencies than high so that many classrooms have an unpleasant "boominess" even though they appear to have plenty of acoustical tile. It requires thick acoutic materials to absorb low frequencies effectively.

Sound transmission from one room or area to another is an important consideration for a good acoustical environment. The isolation is measured under carefully controlled conditions and reported in decibels. A transmission loss of 60 decibels would mean that the wall reduced sound intensity by a million to one. It would require a heavy masonry wall to have such a rating. Ordinary school masonry partitions between classrooms have a rating of about 50 decibels and operable or movable walls of excellent construction and installation have a rating of about 40 decibels. Partitions with a rating of less than 40 decibels may allow annoying sounds to be transmitted, particularly if one teacher is using a sound motion picture and the adjacent one is carrying on a discussion. Low frequencies generally are transmitted more than high ones to produce a "booming" sound in adjacent areas.

Distortion is an indication of how sound is changed during transmission and reproduction. Zero distortion would mean highest fidelity and no measurable change from the original. Five per cent distortion is noticed by most people, but not objectionable. Ten per cent distortion becomes objectionable. High-fidelity equipment is available with less than one per cent distortion.

Another problem with listening to sound is described as "masking," in which some unwanted sound tends to drown out or cover up some of the desired sound. Ventilating systems, fluorescent lamp ballasts, street noises and adjacent classroom noises most often mask school presentations. Sound movie projectors often make so much operational sound that the amplifier must be turned to a high level in order to provide intelligibility, which in turn masks the needed sound in adjacent rooms. More attention needs to be given to designing and installing school equipment with low inherent noise to avoid masking desired sounds.

Masking is measured and reported as the signal to noise ratio in decibels between the average desired sound level and the undesired sound level. This measurement is also used on sound-reproducing equipment such as tape recorders, and the higher the figure the better the machine is in this respect.

Various sound frequencies behave in different ways as they are recorded, amplified, reproduced, bounced off surfaces or transmitted through partitions. The resulting imbalance of frequencies is called filtering. It is possible to filter out either high or low frequencies or any combination of them.

Sounds are converted into electricity in order to be amplified, transmitted over a distance or recorded. The devices for making this energy conversion are called microphones. Microphones differ in conversion method (crystal, dynamic and so forth), in directional characteristic (non-, bi-, unidirectional), impedance (low, high), frequency response and so forth. Microphones are described in greater detail in the section on public address systems.

The amount of electricity produced by a microphone from the sound energy it receives is microscopic. It must be conveyed in special shielded wires or cable into an amplifier to be made strong enough for transmission and conversion into other forms of energy. Amplifiers make use of vacuum tubes or transistors and various other electronic components in order to increase the amplitude of the wave form without changing its frequency or quality. Amplifiers differ in gain (decibels), power output (watts), fidelity (per cent distortion), type of inputs, outputs and so on. Amplifiers are described in greater detail with public address systems.

After a sound has been converted to an amplified electrical signal, it can be transmitted over ordinary wires for long distances. It can then be converted back to sound, radiated by a radio transmitter or recorded.

The electrical signal representing sound is easily recorded by converting it to a wavy groove on a disc, magnetic patterns on tape or an optical pattern on film. Disc recording is considered in detail under record players. Magnetic recording is considered in detail under tape recorders. Optical recording is considered under sound motion picture projectors. The sound pattern on discs, tapes or film can be converted back into electricity by the appropriate machines and again amplified for conversion into sound.

Electrical signals can be converted into sound by loudspeakers and headphones or earphones. These devices differ in power-handling capabilities, efficiency and frequency response. They are considered in more detail with public address systems.

High fidelity is a much used term that does not yet have a definite meaning. It has often been applied by eager salesmen to very ordinary equipment. High fidelity should mean that the reproduced sound is so nearly like the original that a critical listener under ideal listening conditions could not detect the difference. This means that the signal to noise ratio must be great (about 50 dB), the distortion must be low (below 1 per

cent), the frequency response must be wide (50-15,000 Hz), the frequency response must be flat (plus or minus 3 dB, abbreviated ± 3 dB) and the volume range must approximate the original. Apparatus that will meet and exceed these criteria is readily available, but it is not inexpensive. Manufacturers often report only the fidelity of one component of a system, usually the amplifier, and omit equally important items such as the microphone and speaker ordinarily used with it. It is the overall fidelity that is the important thing.

Stereo sound means that an entirely separate signal is picked up, amplified, transmitted or recorded, and changed back to sound for each ear. This system enables a listener to reconstruct in two dimensions the relationships of the original sound sources. If properly done, it is possible to attain a very high degree of realism. The recording microphones and the reproducing loudspeakers are ordinarily placed about 6 feet apart, and the listeners should be about equally distant from the speakers during reproduction. Stereo earphones enable one or more individuals to enjoy excellent stereo sound under a wide variety of conditions. Stereo signals can be recorded on discs or tapes. Most audiovisual equipment outside the music department does not yet make use of stereo sound and is called monophonic.

LIGHT

Light is one of the various forms of energy and the one that makes an impression on our eyes. It is produced by the vibration of atomic and subatomic particles, and the resulting wave motions have extremely short wavelengths, extremely high frequencies and travel at extreme speed. There is some evidence that light also exhibits some characteristics of tiny particles called photons, but the wave theory explains most of the properties associated with audiovisual apparatus and its use. Most natural light is produced by the sun and most projectors use incandescent lamps. Classrooms are often illuminated with fluorescent lamps. Other methods for converting energy into light are infrequently used.

Light waves can be diagramed in a manner similar to sound waves, but they can travel through empty space as well as air. The wavelength or distance between two peaks is so short that it is measured in special units called millimicrons, mμ, or one millionth of a millimeter. One millimeter is about 1/25 inch. The Angstrom unit (A) is also used, and it is one tenth of a millimicron. Light waves belong in a family of electromagnetic waves which includes X-rays, ultraviolet, infrared and radio waves. Part of the electromagnetic spectrum is shown in Table 7.

TABLE 7. Visual Response to Electromagnetic Spectrum

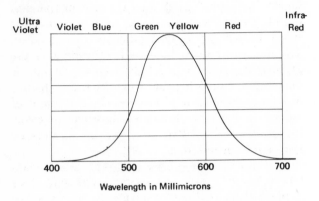

Light travels at a speed of 186,000 miles per second or 300,000,000 meters per second in free or empty space. Its velocity in transparent solids and liquids is considerably slower, and this change in velocity upon entering and leaving them accounts for most optical phenomena. Light travels at approximately the same speed in air as in empty space.

Light rays ordinarily travel in straight lines except when they enter or leave transparent materials that change their velocity. Figure 2.14 shows a source of light and the waves travelling outward in all directions. A ray is one bundle of the waves that passes through a hole or slit and continues in a straight line. As the distance from the source increases, the wave front gets flatter and flatter, and most rays are considered to have flat or plane wave fronts.

The intensity of a light source such as a bulb is measured in candlepower. Previously, one can-

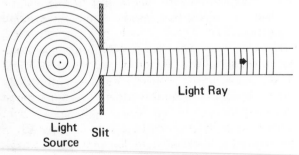

FIGURE 2.14. Light Source and Light Ray

dlepower was the amount of light produced by a standard candle of precise characteristics. A small incandescent lamp is now used as the standard. Automobile headlamps often have 30 candlepower (C.P.) bulbs and an ordinary household 100-watt lamp provides about 100 candlepower. *Lamp* is generally a more technical term for what the layman calls a *bulb*.

The total amount or quantity of light falling on a surface is measured in lumens. A surface of one square foot (12" x 12") located one foot and at right angles to a one candlepower source of light would have one lumen on it. The total amount of light provided by a lighting fixture or projected on a screen is measured in lumens. The total amount of light provided by a one candlepower lamp would be 12.57 lumens. (Lumens equals candlepower times four pi.)

The level of illumination on a surface is measured and reported in footcandles. Pocket meters are available for measuring this quantity directly. The illumination on a surface one foot from a one candlepower source is defined as one footcandle. Since this light would have to cover an area four times as large at a distance of 2 feet, the footcandles would be reduced to one fourth. The formula, often called the inverse square law, is:

$$\text{footcandles} = \frac{\text{candlepower}}{(\text{distance in feet})^2}$$

Using this formula, a 100-watt bulb with 100 candlepower would provide one footcandle of illumination on a desk 10 feet away and one-fourth footcandle on a wall 20 feet away. Due to the inverse square relationship, illumination falls off very rapidly with distance.

Recommended levels of illumination for performing various tasks have been developed at various times. There has been a marked upward trend as lighting systems have been improved and as people have become more critical. Typical recommended levels in footcandles are as follows:

chalkboard, fine sewing	150
drafting, bookkeeping	100
desk work, reading	30-60
corridors, auditorium	10

The human eye can perform under a great range of illumination. Bright sunlight provides about 10,000 footcandles, open shade about 500 fc and bright moonlight about 1/100 fc. It is just barely possible to read a newspaper in 1/10 fc, and this level is often recommended for projection of colored films.

The brightness of a surface depends on the light falling on it and the amount of reflectance from it. Surface brightness is usually measured in foot-lamberts which equal footcandles times per cent reflectance. Some typical reflectance percentages are as follows:

white ceiling	80%
pastel wall	50%
light furniture	40%
chalkboard	20%

Projection screens also depend on reflectance and the brightness is measured in foot-lamberts. It is often recommended that motion picture screens in dark rooms have 16 foot-lamberts on the center of the screen with no film in the projector. When it is necessary to compete with extraneous room illumination, then higher levels would be needed. A powerful classroom filmstrip or 2 by 2 slide projector will put about the same illumination on the screen (30 fc) that the room lights put on the desks. A motion picture projector puts about half as much light on the screen. A matte projection screen (flat white) reflects about 90 per cent of the light striking it.

An opaque substance does not transmit light. A transparent substance transmits most of the light striking it. A translucent substance permits light to go through, but with such scattering of rays that details cannot be distinguished.

The surface brightness of the fixtures providing the illumination is also measured in foot-lamberts. In order to keep the lumens high and the foot-lamberts of the source low enough to avoid unpleasant contrasts, it is necessary to make the light source large. One advantage of fluorescent lamps over incandescent lamps is their large area and lower surface brightness. In practice, most lamps have fixtures to direct light on the task and to avoid high brightness within the usual field of view. For very high levels of illumination (100 fc) it is almost a necessity to have a continuous luminous ceiling.

There are many debates over the superiority of incandescent or fluorescent lamps for classroom illumination. Each has advantages. Incandescent lamps are less expensive, simpler, start instantly and without accessories, make no noise, cause no interference and can be easily dimmed. Fluorescent lamps are much more efficient, last much longer, have lower surface brightness and produce whiter light. Classrooms for a variety of experiences might well be equipped with fluorescents for high-level illumination and a few incandescents for controllable low-level illumination.

Glare is any objectionable brightness contrast. It must not only be controlled in lighting fixtures but from windows and any specular (like a mirror) surface.

The color of light is another important variable. Color is determined by wavelengths, and the human eye responds only to a very narrow band of the electromagnetic spectrum. Table 7 shows the spectrum with the wavelengths and colors labeled. The eye is most sensitive to light energy in the green-yellow area. White light such as daylight is made up of all colors, and black means the absence of light. Color may be determined by the source of light, selective reflectance and absorbing of colors by surfaces and selective transmission of colors by transparent materials.

Incandescent lamps give off very little violet-blue light and much yellow-orange-red light. This is particularly noticeable when color pictures are taken with this type of light and compared with pictures taken in daylight. Projection lamps are operated at high temperatures for short life in order to increase the amount of blue light and improve colors. Arc lamps provide approximately even distribution of colors and intense brightness for theater projection. Fluorescent lamps can be made with various internal coatings to give almost any color balance. Cool-white fluorescents are most common with light approximately placed in the visible region. Light for photographic purposes often has a color temperature rating given in degrees Kelvin. Ordinary projection lamps are at about 3200°K, carbon arcs at 5000°K, blue flash bulbs at 6000°K and blue sky and sun about 6500°K. A candle would have a color temperature around 1500°K.

White light can be easily separated into a band of colors when it strikes a prism at a proper angle. This will be explained later.

Light is converted into other forms for various purposes. In photography light is used to expose and change various substances in order to make a permanent record. In television light is converted into electronic signals which may be recorded, transmitted and converted back to light in a monitor or receiver. A photographic exposure meter and a footcandle meter convert light into electricity which is read by a sensitive meter. In most sound motion pictures, the sound is converted to light variations which expose the sound track which is subsequently illuminated in the projector and the resulting modulated light beam converted to electricity by a photocell system. This is explained in more detail in the section on sound movie projectors. Light is also converted to heat whenever it is absorbed. This is paraticularly important in films and slides when they are exposed to projection lamps. Standard methods have been developed to measure the temperature of film during projection, and systems for keeping it below levels which might damage the film or emulsion have been made. The denser or more opaque the image, the more light will be converted to heat.

When light rays composed of plane wave fronts strike a mirror or similar specular reflecting surface, they are reflected with little loss at an angle equal to the incoming angle. The usual law states that the angle of incidence equals the angle of reflectance. If light strikes a curved surface each ray is assumed to strike a tiny part of the curve which is assumed flat or plane at that spot. A curved reflecting surface may concentrate or disperse a broad bundle of rays. When light rays strike a surface that is rough or matte, then individual rays will be reflected in all directions. Some of the light is also absorbed by such surfaces. Reflection of rays from various surfaces is shown in Figure 2.15.

A projection screen is a special reflecting surface. The idea is to get as much light reflected to the audience as possible without having glare from the reflected projection beam. Problems and solutions are discussed in the section on screens.

Most optical systems include curved reflectors behind the lamps in order to capture that 50 per cent of the light going away from the film and screen and return it to the optical system at the same angle as the emerging rays. The reflector is commonly called spherical because it is part of a sphere.

Some optical systems use large curved reflectors that are parabolic or other nonspherical curves in order to concentrate and direct large amounts of light into special systems. These mirrors are often dichroic, which means they reflect most of the light but not the heat. Reflectors are commonly included inside projection lamps.

Light is not only reflected by mirrors, but by any plane glass surface. Although it may be only about 10 per cent, it is enough to see oneself in a store window, see lights reflected and so forth. This reflected light cuts down the transmitted light, and if several surfaces are involved as in a complex optical system, then special treatment or coating is often used to reduce reflections. Lenses coated to reduce reflection often appear blue or purple by reflected light and the color may not be entirely uniform. Care in cleaning is needed to keep from removing the coating.

Opaque and overhead projectors use mirrors to redirect the image about 90 degrees onto the screen. This mirror must have its reflecting surface on top of the glass rather than protected under it in order to avoid the second image from the glass surface. These mirrors are called first surface mirrors and extreme care must be taken to avoid harming the exposed metallic surface.

When a ray of light falls on the smooth surface of a transparent substance such as water or glass, some of it is reflected as previously explained. The remainder is transmitted through the medium at reduced speed, which will result in bending or refracting if it entered at any angle other than 90 degrees.

Figure 2.16 shows rays entering a substance such as glass at 90 degrees and at about 45 degrees.

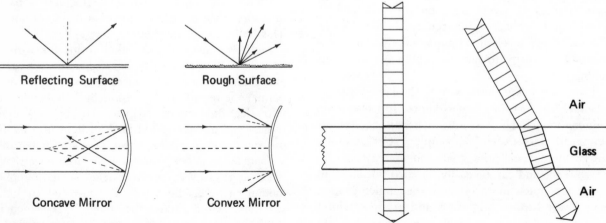

FIGURE 2.15. Reflection of Rays from Various Surfaces

FIGURE 2.16. Light Rays through Glass

When the essentially plane wave fronts enter the glass at 90 degrees, they are slowed down equally at all points and continue at slower speed in a straight line. When they emerge from the glass, they will regain their normal speed, also without change in direction. When the plane wave fronts enter the glass at an angle, one end of each wave front must enter the glass and be slowed down before the other. This results in bending or refraction of the ray. The ray will then continue in a straight line through the glass until it encounters the second surface at which time one end emerges and is increased in speed before the other. A second bending or refraction occurs, and if the two glass surfaces are parallel as in the diagram, then the transmitted ray continues parallel to, but displaced from, the incoming or incident ray. There is also some internal reflection from the second surface so that the transmitted ray has lost some of its intensity.

When light strikes glass with nonparallel surfaces as in a prism, then the passage of a ray can be constructed as in the case above, determining at each surface what happens as plane wave fronts are slowed down or increased in speed. Figure 2.17 shows a ray or beam of light travelling through a

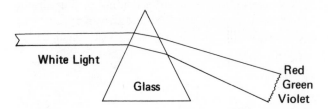

FIGURE 2.17. Light Ray Passing Through Prism

prism. Since different wavelengths or colors travel at different speeds in glass, they will be refracted different amounts and white light will be dispersed into a rainbow of colors.

Different transparent materials transmit light at different speeds so that various degrees of refraction can be selected for desired results.

OPTICS

A projector is a device for making large and bright images on a screen from small slides, films or opaque materials. A projector makes it possible to have the economy, portability and ease of production of small materials combined with the advantages of large images for groups of people to observe and discuss.

Projectors make use of optical systems to recreate large and distant images, and the heart of the optical system is a series of lenses. Lenses are usually considered to be mysterious and beyond comprehension for ordinary people. The design, manufacture and testing of fine lenses is one of the most precise, skilled and complex jobs ever accomplished by man, and not the concern of audiovisual personnel. The selection and use of lenses for audiovisual purposes can be mastered by anyone.

Lenses may be either convex (thicker in the center) or concave (thinner in the center). Most projection lenses are convex, except for internal elements in complex lenses which still behave in combination as convex lenses. Lenses may have one surface flat in which case they are called plano-convex or plano-concave. They may even have combinations of convex and concave surfaces and the result is determined by whether the lens is thicker or thinner in the center. Figure 2.18 shows several lenses.

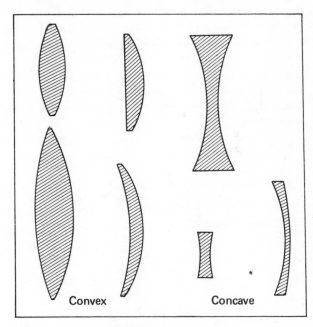

Convex   Concave

FIGURE 2.18. Lenses

In projectors, lenses are used to collect and direct the light onto the slide, film or copy to be projected (condenser lenses), or to recreate the intensely illuminated slide on a distant screen (projection lenses). In either case the lens does its work by bending or refracting light rays.

In Figure 2.19 parallel light rays from a distant source such as the sun enter a plano-convex lens from the flat side. All rays are slowed down but not bent due to the flat surface. As the rays emerge, the one in the center continues without bending, but all others are bent toward the ray that goes through the center. The many rays are concentrated or focused at a spot. The same thing happens with

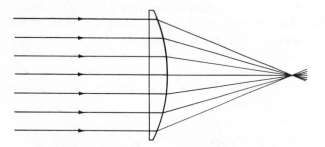

FIGURE 2.19. Focusing of Parallel Rays

a double convex lens, only some bending occurs at each curved surface. This focusing phenomenon can be easily demonstrated with an ordinary reading or magnifying glass and the sun. The rays can be easily concentrated into a spot with such intensity that burning of many objects will occur.

In Figure 2.20 a projection lamp is providing rays which are rapidly diverging according to the inverse square law. A convex lens intercepts some of these rays and concentrates them at the area marked A or brings them to a point at the area marked B.

not used except in combination with more powerful convex lenses.

In Figure 2.21 a pair of small light sources such as colored lamps are used instead of the projection lamp to illustrate the recreation of images on a screen. Lamp a will give off light rays which will be intercepted by the lens and brought back to a point A on a distant screen. Likewise lamp b will give off light rays which are brought to a point B on the screen. Projection with convex lenses always results in inversion of the images on the screen. This is easily corrected by inserting films in the projector upside down and sideways reversed.

Instead of projecting small lampbulbs, it is customary to project brilliantly illuminated slides or film which can be considered as made up of countless small bright areas that are recreated on the screen.

Lenses used in projectors differ in many important parameters or measurable characteristics, but only focal length, aperture and f number are commonly considered in projectors. The fidelity or quality of the detail in the reproduced image is also of importance in considering projection lenses.

The focal length of a lens is the distance between the lens and the point to which it brings parallel rays. This is diagramed in Figure 2.22. In the previous example of using a reading or magnifying glass as a burning glass, the lens was adjusted by trial and error until it was exactly its focal length

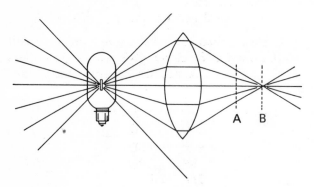

FIGURE 2.20. Projection Lamp and Condenser Lens

Condenser lenses in projectors (usually used in pairs) are designed and installed to capture as much light from the lamp as possible and concentrate it evenly over the slide or film to be projected. A concave mirror or reflector is ordinarily used behind the lamp also.

Concave lenses cause rays to diverge farther apart rather than converge. Concave lenses do not bring rays together to form real images. They are

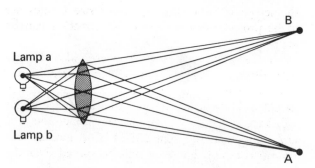

FIGURE 2.21. Inversion of Projected Image

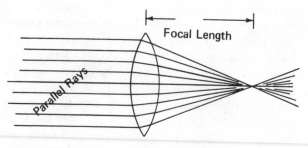

FIGURE 2.22. Focal Length of Convex Lens

from the burning spot. Lenses differ greatly in focal length. Shorter focal lengths make greater magnification. Microscopes have the shortest focal lengths and telescopes have the longest ones. The focal lengths most commonly used in audiovisual projectors are as follows:

8mm movie	1″
16mm movie	2″
35mm filmstrip	4-5″
2 x 2 slide	4-5″
lantern slide	12″
overhead	12-14″
opaque	18″

Different focal lengths are available for most machines. The focal length may be given in millimeters instead of inches. One inch equals 25.4 mm. Focal lengths are usually indicated on the lens housing.

A zoom lens is a variable focal length lens and the range of focal lengths in inches or millimeters is stamped on it. Zoom lenses are more versatile and more expensive than fixed focal lengths. They also may not have the optical quality of fixed lenses.

The focal length of a lens determines the degree of magnification that it will provide. Other things being equal, a 2 inch lens will create an image twice as wide as a 4 inch lens. Several different size images can be created with the same slide and screen placement, simply by changing focal lengths. With a zoom lens an approximate distance can be chosen and then the focal length adjusted so that the screen is just filled. If various size images are to be projected an approximate correction can be made.

The size of the image on the screen also changes with distance. When the projector to screen distance is doubled, the image is twice as large in height and width. An image projected twice as far will have twice the width and four times the area and one-fourth the footcandles. The lumens will remain constant, but they will be spread over the larger area.

The focal length of a projection lens also determines where it must be placed in relation to the slide or film being projected. When focused on a distant screen (50 or more feet distant) the optical center of the lens is placed at exactly its focal length from the slide. This means that lenses must be equipped with tubes, barrels or special supports to locate them at the proper distance. When focusing on a nearby screen, it is necessary to move the lens out farther than its focal length from the slide in order to bring the rays into focus. Lenses are moved in and out on a helical thread or with a rack and pinion. There is some danger in losing the lens out of the mechanism if an attempt is made to focus a very small, bright and nearby image.

There are four linear-size parameters involved in projecting images. Slides or films range from 8-millimeter movies to 10-inch overhead transparencies. Screen sizes range from about 10 inches in study carrels to 12 feet in auditoriums. Focal lengths range from one-half inch to 24 inches. Projection distances range from a few inches to 100 feet. These four linear dimensions are conveniently combined in a variation of the general lens equation as follows:

$$\frac{\text{slide dimension}}{\text{screen dimension}} = \frac{\text{focal length}}{\text{projection distance}}$$

Any three known values can be put in the equation and the fourth one can be computed. It is essential that all values be in the same units, usually inches. Any linear dimensions on the slide and screen could be compared, but horizontal widths are most commonly used. The horizontal widths of commonly projected materials, not including the frames or margins, are as follows:

regular 8mm movie	.17″
super 8mm movie	.21″
16mm movie	.38″
single frame filmstrip	.9″
double frame filmstrip	1.3″
2 x 2 slide	1.3″
lantern slide	3″
overhead and opaque	10″

If it is desired to make a screen image 6 feet wide from a 2- by 2-inch slide with a 5-inch lens at an unknown distance, the equation can be used as follows:

$$\frac{1.3}{6 \times 12} = \frac{5}{D}$$
$$5 \times 6 \times 12 = 1.3D$$
$$360 = 1.3D$$
$$277″ = D$$
$$D = 23 \text{ feet}$$

If a screen is needed for projection of 16mm films with a 2-inch lens from the back of a 30-foot room, it could be determined as follows:

$$\frac{.38″}{S″} = \frac{2″}{30 \times 12}$$
$$2S = .38 \times 30 \times 12$$
$$2S = 136.8$$
$$S = 68.4″ \quad \text{(Next standard screen size is 70″ wide.)}$$

In order to make setting up various pieces of equipment easy under a variety of situations, projection tables have been constructed using the preceding equation and data. These appear on adjacent pages.

The tables contain the information most commonly needed. A single line from the proper table might be copied and left with each machine for ready reference. The body of each table gives the projection distance in feet corresponding to the focal length of the projection lens and the size of the screen that is used. For intermediate focal length lenses, an approximate projection distance can be obtained by interpolation. The tables are based on the horizontal widths given.

The tables are not only useful for determining proper projection distance for a particular screen and focal length lens, but they can also be used for selecting a suitable lens and screen size for certain projection distances. With 16mm motion pictures, if the projection distance is about 40 feet, a 2-inch lens could be used with a 96-inch screen width, or a 3-inch lens could be used with a 60-inch screen, or a 4-inch lens could be used with a 50-inch screen.

Since motion pictures, single frame filmstrips (which are most common) and lantern slides have a rectangular picture with the long dimension horizontal, rectangular screens are indicated. The other machines have square screens indicated. If a screen is going to be used with both rectangular and square pictures the square screen should be specified.

The aperture of a lens is the effective diameter of the largest glass element and it controls the size of the bundle of rays that can pass through. It can be measured with a ruler. A larger aperture with the same focal length means that more light can be gathered and a brighter image created on the screen. Since bright pictures are important, large apertures are very desirable.

However, brightness of picture or light-gathering ability is also controlled by focal length as a simple diagram, Figure 2.23, indicates.

Hence, in determining light-gathering ability of a lens or its ability to project bright pictures, both its focal length and aperture must be considered. The ratio of the two is a convenient way of comparing lenses and is known as the f number.

$$\text{f number} = \frac{\text{focal length}}{\text{aperture}}$$

Both the focal length and the aperture should be in the same units, probably inches. Since it is a ratio, the f number has no units.

Inspection of the formula will show that smaller f numbers mean greater light-gathering ability. Light-gathering ability is often called the speed of a lens because that is what it amounts to in a camera; for example, a lens that gathers a lot of light enables one to take a picture with a short exposure time.

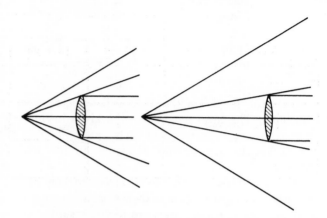

FIGURE 2.23. Light Gathering Ability of Lenses with Same Apertures and Different Focal Lengths

It is important to have lenses with small f numbers on projectors in order to get bright pictures. This is particularly important if the room is not very dark or if the picture must be magnified many diameters. Common projector lenses have the following f numbers:

8mm movie	f/1.4
16mm movie	f/1.6
filmstrip-2x2	f/3.5
overhead-opaque	f/3.5

Of course, the projection lamp, condenser system, distance and screen surface also have a pronounced effect on image brightness.

It is easy to get great speed (low f number) with short focal length lenses, but very difficult and expensive to get long focal length low f number lenses because the aperture needs to be so large. If the focal length is doubled, then the diameter would need to be doubled to maintain the same f number.

There is sometimes a choice of f number projection lenses and projection lamps. Doubling the wattage of the lamp should put approximately twice as much light on the screen. The f numbers are related as their inverse squares so to get twice as much light as an f/3.5 lens would require an f/2.5 lens. To get twice as much light as an f/1.6 lens would require an f/1.25 lens. The faster lens would

PROJECTION TABLES

The body of the table gives the approximate projection distance in feet for the lenses listed in the left column and the screens indicated across the top in inches.

8mm MOVIES (REGULAR 8)

Lens Focal Length	Screen Size					
	12 x 16	18 x 24	30 x 40	37 x 50	45 x 60	52 x 70
1/2″ = 12.5 mm	4	6	10	12	15	17
1″ = 25 mm	8	12	19	24	29	34
1 1/2″ = 38 mm	12	17	29	36	44	51

8mm MOVIES (SUPER 8)

Lens Focal Length	Screen Size					
	12 x 16	18 x 24	30 x 40	37 x 50	45 x 60	52 x 70
1/2″ = 12.5 mm	3	5	8	10	12	14
1″ = 25 mm	6	10	16	20	24	28
1 1/2″ = 38 mm	10	14	24	30	36	41

16mm MOTION PICTURES

Lens Focal Length	Screen Size					
	30 x 40	37 x 50	45 x 60	52 x 70	63 x 84	72 x 96
1″	9	11	13	15	18	21
2″	18	22	26	31	37	42
3″	26	33	40	46	55	63
4″	35	43	53	61	74	84

PROJECTION TABLES

The body of the table gives the approximate projection distance in feet for the lenses listed in the left column and the screens indicated across the top in inches.

SINGLE FRAME FILMSTRIPS

Lens Focal Length	Screen Size					
	30 x 40	37 x 50	45 x 60	52 x 70	63 x 84	72 x 96
4"	15	18	22	26	31	35
5"	18	23	27	32	38	44
7"	26	32	38	45	54	62

DOUBLE FRAME STRIPS AND 2 by 2 SLIDES

Lens Focal Length	Screen Size					
	40 x 40	50 x 50	60 x 60	70 x 70	84 x 84	96 x 96
4"	10	12	15	17	21	24
5"	12	16	19	22	26	30
7"	17	22	26	30	36	42

LANTERN SLIDES

Lens Focal Length	Screen Size					
	30 x 40	37 x 50	45 x 60	52 x 70	63 x 84	72 x 96
10"	11	14	17	19	23	27
12"	13	17	20	23	28	32

OVERHEADS AND OPAQUES

Lens Focal Length	Screen Size					
	40 x 40	50 x 50	60 x 60	70 x 70	84 x 84	96 x 96
14"	5	6	7	8	10	11
18"	6	8	9	11	13	14
22"	7	9	11	13	15	18

cost considerably more compared to the higher wattage lamp but would never need replacing and the heat would be half as great.

Lenses differ in the quality or fidelity of the image they recreate on the screen as well as magnification and brightness. This quality is made up of color fidelity and resolution or sharpness of the image. The defects or imperfections in a reproduced image are called aberrations, and they can only be corrected by very painstaking combinations of several lenses. There are six common aberrations that must be eliminated in high-quality lenses:

1. Spherical—outer rays do not come to same point as central rays. Results in blurred images.
2. Chromatic—images fringed with color due to prism effect of lens.
3. Astigmatism—lines at right angles not equally sharp.
4. Coma—blurring of image edges due to unequal bending.
5. Curvature of field—sharp image is saucer shaped. Only the center or edges can be brought into sharp focus on a flat surface.
6. Distortion—straight lines become curved. Produces a pin-cushion effect from a square.

All these aberrations are present when a single convex lens is used for projection. They can be demonstrated with an ordinary reading glass used to project the image of a low wattage unfrosted bulb on a screen. The combination of these defects results in poor resolving power, definition or sharpness of the image. Resolving power can be determined by projecting precise test patterns under controlled conditions. The result is a number of lines per millimeter that can be distinguished in the center and near the edges. The center resolution is ordinarily much better than the edges.

In order to obtain lenses with low aberration and large apertures, it is necessary to combine several lenses. Most high-quality projection lenses are made of three elements called a triplet as in Figure 2.24. The center lens is concave, but the whole combination behaves as a convex lens. Many overhead projectors for chalkboard substitutes use two element lenses in order to reduce cost for an uncritical use. Overheads for the projection of fine detail should have three element lenses.

The projected image should be uniformly illuminated. A difference of two to one between the center brightness and the corners is not objectionable, but no greater ratio should be tolerated for critical use.

Flare is often observed with inexpensive lenses and transparencies that have opaque areas. It can

be easily demonstrated by punching holes in an opaque sheet and projecting them on an overhead projector with a one or two element lens. The bright spots will have ghosts around them.

Keystone distortion refers to a screen image that is wider at the top than at the bottom due to

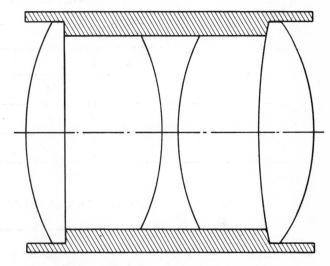

FIGURE 2.24. Three Element Projection Lens

greater projection distance from a low projector to a high screen. It is most pronounced with overhead projection. Figure 2.25 shows keystoning and the usual method for correcting it.

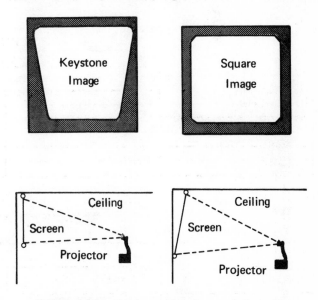

FIGURE 2.25. Keystone Distortion and Correction

PROJECTION LAMPS

In order to project a large and bright image on a distant screen, it is necessary to have an intensely illuminated slide or film in the projector. This illumination is nearly always done with incandescent lamps in conventional audiovisual projectors. Theater projection is generally done with carbon-arc lamps in which direct current arcs between two electrodes produce a brilliant blue-white spot of light. Some auditorium or special purpose projectors make use of xenon lamps which will be briefly discussed later.

Incandescent lamps are familiar because they have been used for general household illumination for so many years. Automobile and portable lantern and flashlight bulbs are also incandescent. Fluorescent lamps are not suitable for projection.

The light from incandescent lamps comes from heating a filament of tungsten wire to about 5000° F by passing electricity through it. It makes no difference whether it is direct or alternating current. In order to keep the tungsten wire from burning up it is necessary to operate it in an atmosphere that contains no oxygen. In practice an atmosphere of nitrogen and argon is usually used at about normal atmospheric pressure. As the lamp is operated, some of the tungsten vaporizes or boils off at such high temperature, and it must be deposited somewhere. When it deposits on the glass walls, it produces the familiar blackening which reduces the light transmitted. When the filament loses enough of its metal, it gets weak and burns out or falls apart. The filament is very strong when cold and very weak when operating near its melting point. Cold tungsten has lower resistance, and hence more current is used when the switch is turned on. Most lamp failures occur at the instant they are switched on. Switching lamps on and off decreases their life somewhat, but not enough to interfere with turning machines on and off to suit educational needs.

All lamps have fuses built into their bases so that an internal short circuit during filament failure will not burn out a fuse in the building circuit.

Although projection lamps are similar to household lamps, they differ in important ways as seen in Table 8.

Household lamps are housed in frosted- or etched-glass envelopes in order to diffuse the light, increase the apparent size of the source and decrease the surface brightness. Projection lamps have clear glass envelopes so that the direct rays can be intercepted and used efficiently in the optical system.

The filament of household lamps is relatively large and strung out while in projection lamps it is concentrated in order to be used efficiently by the optical system.

Household lamps are cooled by direct radiation and convection currents into the surrounding area. Most projection lamps have so much heat to dissi-

TABLE 8. Comparison between Household and Projection Lamps

	HOME	PROJECTION
Enclosure	Frosted	Clear
Filament	Diffuse	Concentrated
Cooling	Natural	Forced
Size	Large	Small
Shape	Arbitrary (A)	Tubular (T)
Glass	Soft	Hard
Base	Screw	Prefocus
Cost	Low (40¢)	High ($4.00)
Life	1000 hrs.	25 hrs.
Watts	25–150	300–1000
Position	Any	Base Down

pate from a small area that fans or blowers must be employed to keep the glass envelopes from melting. These fans are ordinarily on a special switch or thermostat so they can be operated for a moment after the lamp is turned off. With the special switches it is possible to operate the fan without the lamp but not the lamp without the fan.

Projection lamp envelopes are much smaller than household lamps because they must fit into optical systems where space is very important.

Most household lamps have a peculiar shape referred to as arbitrary (A), while projection lamps are usually tubular (T). Other shapes may be used for special purposes.

The glass used in projection lamps is very hard and strong, and it has a high melting point. It is actually very difficult to break the lamp envelope when cold.

Household lamps commonly use a medium screw base about one inch in diameter. This screw base was once used for projection lamps, but it required accurate repositioning of every new lamp in its optical system. All projection lamps now make use of some form of prefocus base or ends that accurately place the filament in the center of the optical system. No adjustments are necessary unless the machine has been dropped or tampered with. Typical prefocus bases are shown in Table 9.

TABLE 9. Prefocus Projection Lamp Bases

| Tru-Beam 2-Pin | Double Contact Bayonet | D.C. Med. Ring (Base Up) | Mogul Bi-Post | Single Contact Bayonet | B&H Small Ring |
| Double Contact Prefocus | Tru-Focus | Medium Prefocus | B&H Large Ring | Mogul Prefocus | Medium Screw |

Courtesy of Sylvania

Household lamps are inexpensive with forty cents being a typical cost. Projection lamps are much more expensive with prices ranging from three to eight dollars. Since lamps are often taxed, a nonprofit institutional purchaser should ask for a tax exemption certificate with the invoice.

The life of household lamps is usually around a thousand hours of continuous burning. Due to the need for much more light from a small source, projection lamps do not last nearly as long. The highest efficiency lamps are rated at ten hours, common

usage lamps at twenty-five hours, and tungsten-halogen lamps at seventy-five hours. No incandescent lamp is guaranteed to burn for its rated life. Rated life merely indicates that a large number of them will average at least the rated life under specified conditions. Some new lamps fail at the instant they are turned on due to gas leakage or a deformed filament from rough handling. In the former case the inside of the bulb will turn white from oxidized tungsten.

Projection lamps consume much larger amounts of electrical energy, measured in watts, than household lamps. Typical projectors use from 150 to 1200 watts with the most common amounts between 300 and 1000. Light output is roughly proportional to watts except in the case of special types designed for extra light and short life, or long life and less light. Most of the electrical energy is converted into heat, which means that the lamps get very hot. The bulb temperature is around 1000° F, which will ignite many things and produce serious burns. Lamps should not be touched until the fan has cooled them for some time. If anything should happen to the fan, or if its intake should be blocked by a piece of paper or film, then the bulb is apt to melt, gradually expand and finally fail. High bulb temperatures under ordinary conditions will cause oil from the skin to fuse to or even combine with the glass so it is a good practice to handle lamps with an improvised mitten of cloth or paper.

Household lamps may be operated in any position, but projection lamps are designed for one position plus or minus 20 degrees. The heavy, concentrated filament must be properly supported in a near-melted condition and the vaporized tungsten should be deposited on a grid or in a part of the bulb where it will not interfere with light transmission. The operating position is often stamped on the lamp and it is most often base down. Projectors should not be focused on the floor or ceiling!

Projection lamps are commonly described in terms of the old or new code. The old code provided the watts, shape, size, base/filament construction and volts in order. A 750T12P-120 lamp consumes 750 watts, has a tubular glass envelope, is 12/8 of an inch maximum diameter, has a medium prefocus base and is designed for operation on 120 volts. The new code for this lamp adopted by the United States of America Standards Institute is DDB. This may be called the USASI or ASI (American Standards Institute) or ASA (American Standards Association) code. The new three-letter code is in standard use, and Table 10 lists the characteristics of some common lamps. The three

TABLE 10. Projection Lamps (Most Popular Types)

ASI CODE	TYPE	WATTS	LIFE	BASE	NOTES
BEC	150B12	150	25	DC Bay	
BEH	150T10TF	150	15	Tru Focus	
BEJ	200B12	200	25	DC Bay	
BRH	1000T5Q	1000	75	RSC	Tungsten Halogen
BVB	30T75C	30	25	SC Bay	
CAL	300T10TF4	300	25	Tru Focus	Internal Reflector
CAR	150T10TF1	150	15	Tru Focus	Internal Reflector
CBJ/CBC	75T8/73	75	50	SC Bay	
CBX/CBS	75T8/107	75	50	DC Bay	
CCM/CHD	200T8DC	200	25	DC Bay	
CEW/CFC	150T8/71	150	25	SC Bay	
CLS/CLG	300T8½/11	300	25	SC Bay	
CLX/CMB	300T8½/12	300	25	DC Bay	
CMV/CMT	300T8½/2SC	300	25	SC Bay	
CTS	IM/T12TF	1000	25	Tru Focus	Internal Reflector
CVX/CVS	200T10/87	200	50	Med Pref	
CWA	750T12TF	750	25	Tru Focus	Internal Reflector
CYC	300T10/SC	300	25	SC Bay	
CYM/CYF	300T10/2SC	300	300	SC Bay	
CYS	1200T12TF	1200	10	Tru Focus	Internal Reflector
CZA/CZB	500T10TF1	500	25	Tru Focus	Internal Reflector
CZX/DAB	500T10P	500	25	Med Pref	
DAH	500T12TF3	500	200	Tru Focus	Internal Reflector
DAK	500T10TF	500	25	Tru Focus	
DAY	500T10TF3	500	30	Tru Focus	
DBT	500T10P/HV	500	25	Med Pref	230 Volts
DCA	150T12TFR/LV	150	15	Tru Focus	Low Volts, Reflector
DCL	150T12TF/D	150	15	Tru Focus	Internal Reflector
DDB	750T12P	750	25	Med Prof	
DDW	750T12/24	750	10	Med Pref	
DDY	750T12/34	750	200	Med Pref	
DEC	750T12/5	750	25	DC Med Ring	Burn Base Up
DEF	150T12TFR/LVD	150	15	Tru Focus	Low Volts, Reflector
DEJ	750T12/3LR	750	25	L Ring	
DEK	500T12TF2/H	500	25	Tru Focus	Internal Reflector
DEP	750T12TF	750	25	Tru Focus	
DFA	150T12TFR	150	15	Tru Focus	Internal Reflector
DFC/DFN	150T12TFR1	150	15	Tru Focus	Internal Reflector
DFD	IM/T12P	1000	10	Med Pref	
DFG	150T12TFR5	150	15	Tru Focus	Internal Reflector
DFK	IM/T12/3LR	1000	10	L Ring	
DFR	500T12/DFR	500	25	Locking 4 Pin	Internal Reflector
DFT	IM/T12/46	1000	25	Med Ref	
DFW	500T12TF2	500	25	Tru Focus	Internal Reflector
DFY	IM/T12/5LR	1000	25	L Ring	
DGA	300T10TF	300	25	Tru Focus	
DFG	500T12TF	500	25	Tru Focus	
DGH	750T12/500	750	500	Med Pref	
DGR	750T12TF2	750	25	Tru Focus	
DGS	IM/T12/1	1000	10	DC Med Ring	
DHN	500T12TF3	500	25	Tru Focus	Internal Reflector
DHT	1200T12/50	1200	10	Med Pref	
DJL	150T12TFR3	150	15	Tru Focus	Internal Reflector
DKK	750T12P/HV	750	25	Med Pref	230 Volts
DKM	250T14TFR/LVD/1	250	25	Tru Focus	Internal Reflector
DKR	150T14TFR/LVD1	150	15	Tru Focus	Low Volts, Reflector
DLH	250T14TFR	250	15	Tru Focus	Horizontal, Reflector
DLR	250T14TFR/LVD	250	10	Tru Focus	Internal Reflector
DLS/DHX	150T14TFR/LVD	150	15	Tru Focus	Low Volts, Reflector
DMH	250T14/TFR/D1	250	15	Tru Focus	Internal Reflector
DMS	500T20	500	50	Med Screw	
DMX	500T20P	500	50	Med Pref	
DRC/DRB	IM/T20/13P	1000	50	Med Pref	
DRS	IM/T20/MP	1000	25	Med Pref	
DVY	650QG7/TP	650	50	2 Pin	Tungsten Halogen
DWY	650T4/CL	650	25	RSC	Tungsten Halogen
DYP	600QG7/TP	600	75	2 Button	Tungsten Halogen
DYS	600QG7/TP2	600	75	2 Pin	Tungsten Halogen
DYV	600QG6/TP	600	75	2 Pin	Tungsten Halogen
FAD	650T4/CL1	650	100	RSC	Tungsten Halogen
FAL	400T4½/Q	400	75	RSC	Tungsten Halogen
FCB	600T4/Q	600	75	RSC	Tungsten Halogen
FCS	150T5/2	150	50	2 Pin	Tungsten Halogen
FFJ	600T4/120	600	75	RSC-2	Tungsten Halogen

letters are completely arbitrary, which means that a DDB lamp may bear no close resemblance to a DDA or DDC lamp.

Some lamps are removed from their sockets by pulling, and some by turning counterclockwise and pulling. Some are ejected by levers which may pop the lamp up into the air. Some machines have devices for moving a fresh lamp into operating position without opening the machine. Careful inspection and trial may be necessary in order to change an unknown lamp. If the code can be read on top of the lamp, then the chart (Table 10) will indicate the base construction. If the lamp must be turned during insertion, it is always clockwise. When properly inserted the lamp filament will always be at right angles to the optical axis through the lenses, and the reflector, if included, will be away from the lenses.

Many newer projection lamps are described as tungsten-halogen types. They were called quartz-iodine when first developed. They have several advantages that make them very attractive as new projector models are developed. Tungsten-halogen lamps are usually much smaller in size than ordinary projection lamps, which makes more compact equipment possible. The halogen (chlorine-iodine-bromine) inside the envelope at high temperature combines with the evaporated tungsten and redeposits it on the filament. This action keeps the envelope clean for practically constant light output and prolongs lamp life about three times. A typical projection lamp will be delivering only about 75 per cent initial light at twenty-five hours whereas a tungsten-halogen lamp will be delivering 95 per cent initial light output at seventy-five hours. The lamp envelope should never be touched with the bare hands or it may soon fail when heated up.

Projection lamps are very sensitive to changes in applied voltage. They will provide the rated life and illumination only when the voltage stamped on the lamp agrees with the applied voltage which may be checked with a voltmeter (usually AC). If a lamp marked 115 volts is operated on 130 volts, its life will be about 20 per cent of normal, and it will provide about 150 per cent of normal light. If a 115-volt lamp is operated on 100 volts, its life will be about 600 per cent, and it will provide about 60 per cent of normal light. There may be circumstances when very long life or increased illumination are deliberately sought. Most projection lamps are now designed for 120 volts and most places have close to this voltage. If not, then special order lamps may be obtained for 105, 110, 115, 125 or 130 volts. Another solution is to get the power

company to change transformer taps. A last resort is to use a variable autotransformer to change the voltage at the machine. A line voltage monitor is shown in Figure 2.26.

Special high efficiency lamps designed for 8 to 30 volts are used in some projectors which have built-in transformers. These low voltage lamps provide considerably higher lumens on the screen per watt, and transformer taps may provide various levels of illumination and life.

FIGURE 2.26. Line Voltage Monitor

Courtesy RCA

Projection lamps traditionally contained only a filament, and the condensing system for concentrating light on the film was entirely external. The next development was to put a small proximity reflector inside the lamp. In some lamps there is now a large reflector which nearly surrounds the filament, and no condenser lenses are used. This large reflector may be dichroic, which means it reflects most of the light but very little of the heat. A few projection lamps have both an internal reflector and internal condensing lens.

The light produced by a filament is roughly proportional to the watts put into it, but the light (measured in lumens) put on the screen is determined also by the efficiency of the whole optical system. There are 150-watt systems putting as much light on the screen as other 750-watt systems. The USASI has established a standard method (USASI Standard Z38.7.5) for determining uniformity and amount of illumination on a screen. When brightness of projected image is important it should be specified in lumens, and the lowest wattage providing it would mean less heat and power consumption. Typical 16mm movie projectors put 750 lumens on the screen, and overheads put about 1800 lumens on the screen.

When it is necessary to obtain very high levels of screen lumens with 16mm films or 2 by 2 inch slides, a xenon or quartz short arc lamp may be used. The light has a higher color temperature than incandescent sources so that pictures appear more blue-white. Several times as much light can be put on the screen to accommodate large audiences or for use in partially darkened rooms. The lamps have no filament. A very small but brilliant arc is established in a small tube with a special and heavy power supply. The cost is much above ordinary systems.

Part II

Mediaware

chapter 3

Basic Optical Systems

Background

A projector is essentially a device for producing an enlarged image on a screen by optical means. It permits the production, reproduction, transportation and storage of small images such as slides that can easily be projected into large images when needed for group study. There is nothing recent in the process except brighter and more faithful results. A simple demonstration will show the essentials for projecting images. A small, clear (not frosted glass) electric lamp is set up and lighted. A simple magnifying or reading glass is held near it. The lens is then moved back and forth by trial and error a few inches from the lamp while watching the wall and a bright inverted image of the lamp filament will be seen. This is diagramed in Figure 3.1.

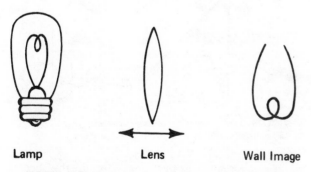

Lamp   **Lens**   **Wall Image**

FIGURE 3.1. Lens Used to Create Projected Image

In this demonstration a lamp filament was used as an image to be enlarged and recreated on the wall by the lens because the filament is very bright. Ordinarily the image to be recreated is a picture or diagram on film, glass or paper, and it must be brightly illuminated by an incandescent projection lamp. A simple setup in a darkened room can be used to show this. See Figure 3.2. It will be noticed

that the screen image is neither very bright or evenly illuminated.

The slide or film should be placed very close to the lamp in order to catch as much light as possible, and far away to keep it cool and evenly illuminated. The dilemma can be solved through the use of a spherical reflector behind the lamp and a condenser lens or lenses between the lamp and slide. The reflector will catch much of the light in back of the lamp which would otherwise be wasted and return it through the lamp filament to combine with the front rays. The rays in front of the lamp are diverging rapidly so that much more illumination would reach the center of the slide than the edges. A system of lenses can change diverging rays to converging rays, and intense even illumination results. The positions of reflector, lamp, condensing lenses and slide must all be carefully fixed to give the best illumination of the slide. A piece of heat

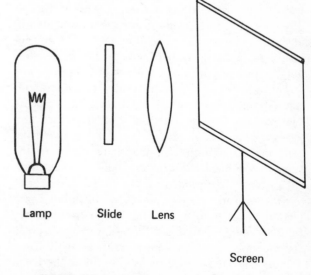

Lamp   **Slide**   **Lens**

Screen

FIGURE 3.2. Lens Used to Project Slide Image

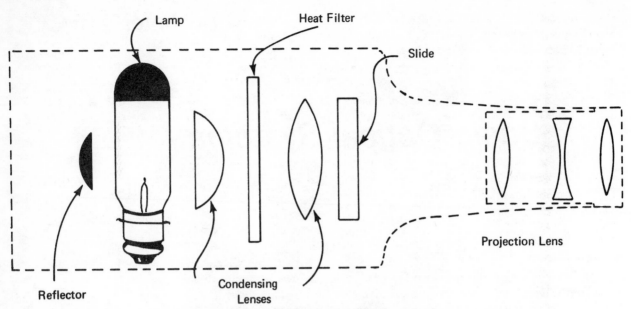

FIGURE 3.3. Straight—Through Projection System

filtering glass is often placed between the lamp and the slide to reduce the transmitted heat. This glass is identified by plane surfaces and a green or blue tint. Some condensing lenses are made of heat filtering glass. A small fan is usually used to cool the lamp, machine and slide. The fan is often wired to a thermostat or separate switch so that it can be left running a minute after the lamp is turned off to cool the lamp and mechanism.

The reading glass must be replaced with a system of precise lenses called a projection lens in order to get an image that is sharp, bright and free from color separation. It is placed approximately at its focal length (stamped on the lens barrel) from the slide, but the distance must be variable in order to get a sharp image at various distances. Every time the screen is moved nearer or farther away, the projection lens must be adjusted to recreate a sharp image. Different focal lengths provide different degrees of magnification.

The system just described is often called straight-through projection and is used in most filmstrip, two by two, lantern slide and motion picture projectors. The optical axis is one straight line from reflector through the lamp and all lenses to the screen. A typical filmstrip or slide projector optical system is diagramed in Figure 3.3.

An overhead projector makes use of indirect projection in order to have a horizontal transparency for easy manipulation and the usual vertical screen for viewing. If a base-down lamp is used as in Figure 3.4, then a mirror is used to direct the beam of light upward after it has gone through one con-

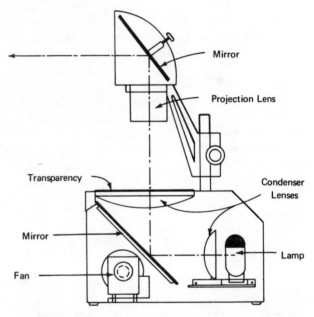

FIGURE 3.4. Indirect Optical System for Overhead Projection

densing lens and before it goes through the second condensing lens just under the transparency. Since the second condensing lens must be very large in order to cover the usual 10- by 10-inch transparency aperture, it is often a fresnel type made of one or two pieces of plastic. The essential curved surfaces have been split up into rings and the thick area between them, which serves no purpose, eliminated. A comparison between a plano-convex lens and its fresnel equivalent is shown in Figure 3.5. Tungsten-halogen lamps operate horizontally and are usually

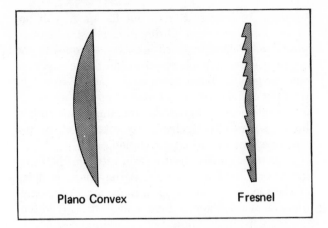

FIGURE 3.5. Equivalent Condenser Lenses

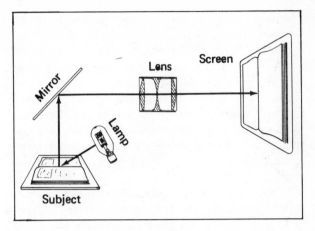

FIGURE 3.7. Opaque Projection

directly under the transparency and fresnel lens as shown in Figure 3.6.

Projection of paper prints requires reflected or opaque projection. Since it is necessary to create

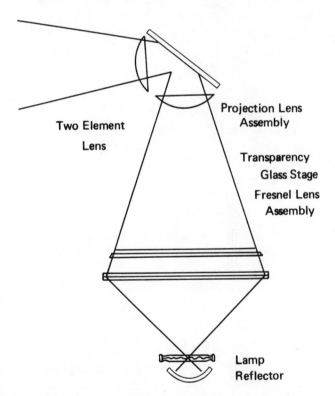

FIGURE 3.6. Simplified Overhead Projection System

the projected image from the light reflected and scattered from a piece of paper, the resulting image is only a small fraction of the brightness obtained with direct projection. The essentials of opaque projection are shown in Figure 3.7.

Selection

The object of the basic optical system is to re-create a bright and faithful image of the slide, film or paper print on the screen with a minimum amount of heat, noise and inconvenience. Selection factors for each machine will be considered separately as they are studied in later sections.

Operation

The assignment sheet for the operation of a typical projector optical system is included at the end of this section.

Maintenance

Optical systems need to be cleaned regularly because the fan, or convection currents circulate dirt through the machine.

Glass condenser lenses can be cleaned with a dry or damp cloth or with soap and water. Oil can be cleaned off with lighter fluid or dry cleaners. Plastic fresnel lenses require more care in order to avoid scratches, and only water or a liquid cleaner specifically made for use on plastics and a soft cloth are recommended. The two parts of the plastic fresnel sandwich often used in overheads should not be separated. It is essential that condenser lenses with unequally curved surfaces, which are usual, be put back in the machine in the correct order and with the correct surface facing the screen. If improperly assembled, the lens holder may not fit, a lens may be too close to the lamp or slide, and an unevenly illuminated image will result. If no diagram is provided in the machine, trial and error may be required with momentary use of the lamp to straighten out an improperly assembled machine.

Projection lamps other than tungsten-halogen types can be cleaned like glass condenser lenses.

The small tungsten-halogen types should never be touched with the bare hands or body oil will tend to destroy them. These lamps must be wrapped with cloth or paper during handling. A spare lamp should accompany each machine.

The mirror behind the lamp, if used, can be wiped out with a dry or damp cloth. The plane mirrors used between the transparency or print and the projection lenses in overheads and opaques require special care in cleaning which is explained in the sections on these machines.

Projection lenses are ordinarily coated to reduce reflections, so they must be gently cleaned. A special tissue called lens tissue is specially made for this purpose. It works very well on a lens surface that has been fogged with the breath. If lens tissue is not available, a well-laundered handkerchief can be used. If a scratch should appear on the lens, the light will be scattered, but its image will not appear on the screen. Ordinarily, only the two exposed outer lens surfaces are cleaned outside the repair station. Dirt observed inside the lens system cuts down on illumination and scatters the light rays but specks are not recreated on the screen. If a lens element is loose, the retaining ring should be very carefully tightened by rotating it, or the machine sent to the repair station.

Fans occasionally need a drop of oil in the bearing at each end of the shaft if they become noisy or slow to start. Excess oil must be avoided. Sometimes a metal blade is bent and needs straightening to keep the machine from vibrating.

Various exposed parts of the machine such as feet and switches may get loose. They may be tightened with simple tools or sent for repair.

Assignment I. Checked _____

Name_____

Date_____

BASIC OPTICAL SYSTEM

Set up a straight-through lantern slide projector or miniature slide and/or filmstrip projector with the appropriate slide or filmstrip and focus the image on a wall or screen by moving the projection lens in or out. Experiment with inserting the slide or strip to get a true reading and erect image.

Open the machine to get at the optical parts. This will ordinarily require no tools since these parts must be cleaned regularly. Handle optical parts carefully by the edges to avoid finger marks. The location and orientation of any parts removed must be carefully noted so that they can be returned exactly as they were.

Caution! High-powered projection lamps get dangerously hot. Leave the lamp on just long enough to make the observations, and let it cool before touching it, or adjacent parts.

1. Make and model of projector used _____

2. Width and height of image on film or slide_____

3. Projection lamp watts_____ , volts_____ , three letter code _____

4. How is the projection lamp removed?_____

5. Diameter of reflector in inches _____

6. Number of condenser lenses_____ , diameter of largest one_____

7. Is a heat filter used? (describe) _____

8. Is a fan used?_____ Can it be operated without the lamp?_____

9. Focal length of projection lens _____

10. Aperture of projection lens_____ (diameter of largest glass element)_____

11. Method of focusing the projection lens_____

12. Deliberately simulate dirt on the lens by placing a pencil across the front of the lens. Describe the result on the screen _____

13. Make a diagram to show how the image is inverted by the projection lens.

14. Remove or momentarily cover the reflector behind the lamp with something noncombustible. Be careful not to get it in the fan. How much does the reflector contribute to screen brightness?

15. Remove one condenser lens and *momentarily* turn on the lamp. Describe the screen image obtained without it.

16. What is the minimum distance between projector and screen at which a sharp image can be obtained? In checking this, be very careful not to drop the projection lens out of its housing. (It may be easier to move the screen than the projector.)

17. What is the minimum horizontal dimension of the projected image?_____What happens to the horizontal dimension and area of the image on the screen or wall any time the projection distance is doubled?

18. Make a cross-section diagram of the complete optical system used in this machine. Label each part.

chapter 4

Standard Lantern Slide Projectors

Background

The standard lantern slide projector is the oldest type of audiovisual equipment still in regular use. Practical apparatus for recreating an enlarged version of an image from a glass plate to some form of screen has been known for two hundred years. During the first quarter of this century the lantern slide and its projector were the most common visual devices in use. Many of these old machines can still be found, and some of them are still in use.

About fifty years ago the Keystone View Company prepared and sold a comprehensive set of six hundred lantern slides to illustrate most phases of the curriculum. A complete teacher's guide in book form and individual cards with each slide made their use easy and effective. Five-hundred watt projectors could be used in rooms with little light control. Both amateur and professional photographers made lantern slides to supplement the commercial ones.

The lantern slide got its name from the "magic lantern" which seemed to provide pictures on a screen with magic, and some form of a lantern for light. Oil burning models were replaced by gas and then by incandescent lamps which were gradually improved. The current models bear a striking resemblance to the models of fifty years ago.

The lantern slide has been standardized for many years at 3 1/4 by 4 inches overall. (It is often called 3 by 4.) The typical slide consists of a photographic positive (direct reading) image on glass, a mat or mask of paper to provide an accurate border and a place to print identification and explanatory material, a second piece of glass called a cover glass to protect the photographic emulsion, and a strip of gummed tape around the outside to bind and protect the sandwich. The outside surfaces can be cleaned as necessary without touching the image or printing. Although the outside dimensions are standard, the opening in the mask varies greatly. A common opening is 2 3/4 inches high and 3 inches across, and the projection tables are based on this projected opening. Lantern slides, even of tall objects like trees, must always have the long dimension horizontal, although a vertical mask can be put within the slide for special effects.

Handmade or school-made nonphotographic lantern slides have been made for many years by teachers and students. Some pencils and inks will work directly on glass for projection. Silhouettes can be cut out and placed between two cover glasses. Small and thin natural objects such as leaves can be mounted and projected. Typewriting can be done on cellophane, and special kits called Radiomats make it very easy and effective. Glass that has been etched or roughened can be marked with ordinary black or transparent colored pencils, and prepared glass for this purpose is available. However, the etched glass scatters much of the light aimed by the condensing lenses into the projection lenses so that screen illumination is much decreased and a dark room is required. Most of the new felt or nylon tip pens for use on the overhead projector will work very well on lantern slide glass. Many other recently developed techniques for overhead projection can be used with this machine. Heavy plastic can be used instead of glass. Some materials too small for proper overhead projection will be enlarged enough by the lantern slide projector to be useful. Details on local transparency preparation are given in another section. Photographic lantern slides are still made on coated glass plates through regular darkroom techniques. Ektachrome or other full color sheets in this size can be exposed in a large camera and projected with excellent results.

The lantern slide projector is one of the simplest machines. The idea is to illuminate an image on glass or plastic very intensely by an incandescent lamp, reflector and condenser lenses and then recreate the image in enlarged form on a distant screen by a combination of lenses called the projection lens. A typical machine is shown in Figure 4.1.

Selection

Lantern slide projection was once commonly combined with opaque projection in a dual purpose unit. These are now museum pieces. Lantern slide attachments are available for some overhead projectors. The best arrangement seems to be to have a separate machine for this one purpose.

Lantern slide projectors differ in many ways, but most importantly in screen illumination, magnification, cooling system and accessories. Convenience and appearance may also be important.

Screen illumination is determined by the projection lamp light output in lumens, the efficiency with which the reflector and condensing lenses pass the light through the transparency and into the projection lens, and the f number or light-gathering ability of the projection lens. Illumination is measured with no slide in the projector. Screen brightness also varies inversely with screen area and distance.

The lumens delivered by a lamp are closely related to the wattage of the lamp unless some of the new specialized lamps are used. Most lantern slide projectors employ 500-watt T20 (tubular, 2 1/2 inch diameter) lamps. These lamps are supposed to have fan cooling, but many of them do not, and instead depend on a large and open box for cooling. A plane (flat) tinted glass heat filter may sometimes be positioned between the lamp and slide in order to reduce the heating of the slide. Auditorium models may use a 750-watt lamp which must have a fan to cool both it and the slide.

The reflector and condensing lenses used on lantern slide projectors are large, approximately 4 inches in diameter, in order to cover the diagonal of the slide with even and intense light. There are usually two convex lenses. The focal length and hence thickness of the condensing lenses must be matched to the focal length of the projection lens in order to have the cone of light through the slide exactly fit the projection lens.

Lenses for projecting lantern slides come in a variety of f numbers for collecting light and focusing it on the screen. The lower the f number, the more light is gathered and an f3.5 lens gathers twice as much light as an f5 lens. Lower f number lenses are often called "high speed" because a similar lens

FIGURE 4.1. Typical Lantern Slide Propector

Courtesy American Optical Company

in a camera would permit pictures to be taken with a short exposure. Low f number lenses are effective only if matched to condenser lenses that will fill the entire diameter of the projection lenses. Increasing the diameter (decreasing f number) of the projection lens does no good unless it intercepts more light. Lower f number lenses increase rapidly in cost, particularly as the focal length is increased. Common f numbers for lantern slide projectors vary between f3.5 and f10.

The focal length of lantern slide projection lenses varies between approximately 6 and 24 inches. The longer lengths are needed in order to project from the back of an auditorium to a distant screen of reasonable size. Even a 24-inch lens will produce an image 15 feet wide at 100 feet. Since long focal lengths have f numbers that are high unless the diameter is prohibitively large, the auditorium must be dark for their use.

An important reason for the lack of advancement in the design of lantern slide projectors is the large size of the optics required.

Fans and heat-absorbing glass may be used in 500-watt models and usually used in 750-watt models. Fans should be checked for objectional noise and the direction of exhaust air.

In addition to the various condenser-projection lens combinations, other accessories and convenience features may be offered.

Slide carriers to hold other than 3 1/4 by 4-inch slides may be desirable. These are available for 3 1/4 by 3 1/4 (British), 2 3/4 by 2 3/4 and 2 1/4 by 2 1/4 (intermediate size camera enthusiasts) and 2 by 2 (35mm and other miniature cameras). The two larger sizes use the same lens systems. Unless the smaller slides are used with accessory lenses, there will be a large loss of light, since most of it will be blocked by the opaque areas of the slide and slide carrier. Before purchasing such accessory lenses, they should be compared with the cost of a separate machine designed for the small slides. Automatic or semiautomatic slide changers are seldom available or used except in special television studios or automated multimedia auditoriums.

Two lantern slide projectors can be accurately positioned for alternate slide projection with a pair of iris diaphragms or light dimmers. This scheme has been used for many years to give the appearance of one slide dissolving into another. The slide is changed only when that projector is not projecting. A pair of the small silicon-controlled rectifiers (SCR) used in house light dimming would do the job effectively.

Operation (assignment sheet at end of section)

The machine should be unpacked and set on a firm table or stand and the lantern slide holder or carrier secured in place with a thumbscrew or other locking device. The carrier allows either of two slides to be easily and quickly put in the correct position for projection. The reason for this is to present a rapid sequence of pictures with no blank screen between them. Two slides should be inserted for trial.

There is some method of extending the part of the machine that holds the projection lens so that the lens will be approximately its focal length (stamped on the lens) from the slide. A plastic or fabric bellows is most often used. When focusing on a distant screen, the center of the projection lens will be exactly its focal length away from the slide. When focusing on a nearby screen, it is necessary to extend the lens out beyond its focal length from the slide. Thumbscrews are often used to secure extension rods and the lens in any desired position. After the lens is approximately positioned, the lamp should be turned on and a picture roughly focused on a screen or convenient wall by moving the lens in and out. Fine focus is obtained by revolving the lens on a spiral thread or rotating a knob.

Since projection lenses invert and make mirror images, the slide must be inserted in the carrier upside down and backwards. To help projectionists, a thumb mark is often placed on each slide in the upper right-hand corner when the projectionist is behind the machine facing the screen. A right-handed person would insert slides with his thumb on the mark. To read the slide correctly without projection, the thumb mark would be in the lower left corner.

The size of the projected image depends on the size of the image on the slide, the focal length of the projection lens and the distance between the projector and the screen. The relationship is as follows:

$$\frac{\text{slide dimension}}{\text{screen dimension}} = \frac{\text{focal length}}{\text{projection distance}}$$

All dimensions should be kept in the same units, probably inches.

The machine has an elevation adjustment to get the picture just the right height on the screen. It may also have a leveling device to rock the machine sideways and level the picture on the screen.

Maintenance

Lantern slide projectors require little maintenance because they are relatively simple and rugged.

The lamp, reflector and condenser lenses should be cleaned regularly with a dry or damp cloth, or even removed from the machine and washed with soap and water. The projection lens should be carefully checked for tight retaining rings and cleaned as described under lenses. The individual lens elements should not be removed except under technical supervision.

All adjustable and removable parts should be kept tight enough so that they will not fall off and get lost.

If the screen image is not evenly illuminated from top bottom or side to side when a normal slide is properly inserted and focused, the optical system needs to be checked. The reflector, lamp filament, condenser lenses, slide and projection lenses must be exactly on one straight line called the optical axis. The most common problem is to have the lamp socket pushed down or up in its mooring due to incorrect lamp replacement. The base may need to be loosened, positioned and tightened. Since the lamp gets very hot, it should be turned on only for very brief periods during the adjustment. The lamp socket may also have been rotated so that the filament is no longer at exact right angles to the optical axis. This is also corrected by loosening and repositioning the lamp. One or more of the condensing lenses may not be seated properly in its housing.

If the screen image is considerably brighter in the center or in the corners, then the projection lamp must be moved forward or backward, or the condensing lenses have been improperly inserted. The lamp socket is often on tracks for forward and backward movement with a screw to lock it in position. If no diagram of the condensing lenses is provided and they are not labeled, then trial and error will be needed to find the correct position and orientation of each lens with brief use of the lamp. With two lenses and two sides for each, there are only eight possible ways and only one way is correct if the lenses are dissimilar, which is the usual case.

If a fan is used, it may have oil holes which need a drop of oil occasionally. Excess oiling will result in a film on optical surfaces.

If the power cord has screw terminals, they should be tightened. Some machines may have a power (on-off) switch in the cord that periodically needs to have screws tightened.

Assignment II.   Checked _____

Name _____

Date _____

STANDARD LANTERN SLIDE PROJECTOR

1. Make and model of machine _____ list price _____

2. Power requirements: volts _____ , amps _____ , watts _____ , AC? _____ DC? _____

3. Does it have a fan? _____ heat filter? _____

4. Projection lamp data: watts _____ , shape _____ , size _____ , base _____ , volts _____
    three letter code _____

5. How is the slide carrier held in place for use? _____

6. How is the projector elevated? _____

7. How is the projector leveled? _____

8. Diagram a slide with a diagram, title and thumb mark correctly placed for projection.

9. Roughly focus the slide on a wall or screen by extending the projection lens, and then sharply with
    the fine focus device. Describe the fine focusing arrangement. _____
    _____

10. Projection lens: focal length _____ aperture _____ , f number _____

11. At what distance will this machine just fill a 37″ x 50″ screen? _____
    by computation _____ from table _____

12. Is there an adaptor for other size slides? _____ what size(s)? _____

13. List any other accessories or special features.

chapter 5

Projection Screens

Background

The purpose of projection is to recreate an enlarged image of the slide in the projector on some distant surface. Actually no special device called a screen is essential since an image can be focused on any surface that is not totally black or specular (like a mirror). Images for special effects have even been focused on water sprays, rocks and people.

In order to obtain good projected images under usual classroom conditions, a special surface mounted in some special way called a projection screen is usually employed.

The traditional screen surface is described as mat or matte white which means that it is white and without luster or gloss. The surface should also be made of such fine or small grains or particles that they will not be visible at any ordinary viewing distance. Mat surfaces have been available for many years and they are still used in most theaters and recommended for many school applications. A good mat screen surface will reflect approximately 85 per cent of the light falling on it with essentially even distribution in all horizontal and vertical directions. In a dark room a projector without a transparency can be focused on a mat screen and the floor, walls and ceiling near it will be observed to be evenly and effectively illuminated by the light scattered and reflected from the minute particles of the mat surface. All persons in the audience will see an equally bright image.

White plaster or flat white paint on a wall makes a convenient and effective mat screen in classrooms that have good light control or powerful projectors.

The mat surface screen, on the other hand, is very wasteful since all the light reflected on the walls, ceiling and floor is of no use to the audience, and when reflected from white ceilings and walls it even adds to the room illumination and competes with the screen image. With high lumen output projectors, such as overheads, or in well-darkened rooms, the low efficiency of the mat screen in directing light to the audience is not important, and the surface is ideal. In other cases a directional screen may be more desirable.

Ideal projection conditions are seldom encountered in classrooms. The darkest part of a projected image is actually a white screen surface that is dark only due to the absence of light or in contrast to a much greater intensity of light provided by the lamp. In poorly darkened rooms, it may be difficult or uncomfortable to attain a ratio of brightnesses that will produce the proper illusion of light and shadow on the screen.

In order to deliver more of the light from the screen to an audience concentrated in a very limited vertical and horizontal angle measured from the line between the projector and center of the screen, special directional screen surfaces have been developed. The idea is to deliver a bright and uniform image to all of the audience and no light anywhere else. It is also desirable to avoid reflecting light from any source except the projector to the audience. Several directional screen surfaces have been developed.

The first directional surface was described as silver but was actually aluminum flakes applied to a surface. Some of these are still used, particularly for three-dimensional pictures produced by polarized light and twin projectors. The silver screen is about half way between a mirror and a mat surface. It can be made to deliver large amounts of light to a limited area. There is always the danger with this screen, however, that there will be direct reflection of the projection beam to some areas making what are called hot spots. The area that is very bright to one viewer may be very dark to another. Such areas are a particular problem from a surface that is not absolutely flat.

The most used classroom screen surface over the past thirty years has been the glass-beaded type. This surface has millions of tiny spherical glass beads applied to the surface of the screen fabric. When light from the projector hits these beads, it is reflected very efficiently back toward the projector and very inefficiently in all other directions. Since the audience is normally seated around the projector, this selective reflection can be very desirable. Many highway signs are surfaced with glass beads to reflect automobile headlights back toward their source and the driver directly behind them. The reflection curves for typical mat and beaded screens are plotted in Figure 5.1. It should be noted that a person seated along the line of projection would observe an image on a beaded screen several

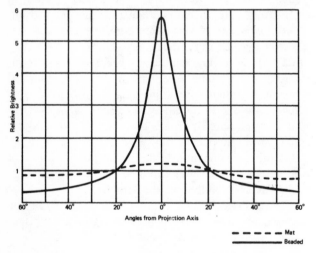

FIGURE 5.1. Mat and Beaded Screens at Various Angles

times as bright as on a mat screen, but at an angle of about 20 degrees either side of the line of projection the two surfaces would appear equally bright. At an angle of 40 degrees either side of the line of projection the mat surface would appear twice as bright as the beaded. Most beaded screens have been sold to people who were standing or seated beside the projector. Most observers are not so fortunate. Beaded screens are recommended for narrow rooms or other situations in which the entire audience can be moved into a cone around the line of projection. Stray light from behind the projector will also be reflected to the audience and stray light from other angles will be attenuated. Beaded surfaces are impossible to clean, and any rough handling may rub off the beads in certain areas so that different degrees of

reflection result. Beadless areas will appear dark to people seated along the line of projection and will not be noticed by people out at an angle. A screen surface not absolutely flat may also have objectional light and dark areas. When highly directional screens are used, it is important for the person in charge to check the viewing in various areas. Viewers in one part of the room may have a very poor image, and offer no objection. The individual beads will be visible to a person seated close to a screen and give a grainy or sparkling appearance.

In order to provide a more uniform image over a wide horizontal angle and a narrow vertical angle, which includes most audiences, the lenticular screen was developed. Lenticular means having the form of a lens since a lenticle is a small lens. This screen surface is more accurately described as a very large number of very small cylindrical silvered mirrors to disperse the projected image over a pie-shaped audience area. The usual construction consists of silvered (aluminum) particles on an embossed plastic sheet. In order to avoid hot spots from an uneven surface, lenticular screens are either attached to rigid backs or provided with tensioning devices. As with beaded surfaces, the image quality from all student angles should be checked.

The most recent screen surface has such directional characteristics that it has been called a sun screen. It is made by applying a specially treated aluminum foil to a concave and rigid back. It is very effective in reflecting light from the projector back uniformly to an audience seated in a rather narrow area around the projector. For this area, the light is about twelve times as bright as from a mat screen. Light from a direction other than the projector will not go to the audience. The screen when observed from outside the intended viewing area will appear dark. This screen when properly used will permit small images to be viewed satisfactorily in lighted rooms or low-power projectors to be used in darkened areas. The screen must be rigid and carefully placed for optimum results.

The optimum relationship between audience seating and the screen is shown in Figure 5.2. W is the width of the screen. The wider angle is for mat and the narrower angle is for directional screens.

All the screens considered so far are designed to reflect the images with the projector and audience on the same side of the screen. This is called front projection. It has the disadvantages of placing the machine and its distractions among the audience, unless a booth is used, and making a shadow on the screen if anyone or anything gets in the line of projection.

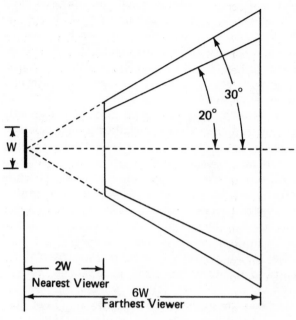

FIGURE 5.2. Screen—Audience Relationships

Translucent screens are used between the projector and the audience and several arrangements provide certain advantages. One of these arrangements is diagramed in Figure 5.3. The earliest translucent screens were made by wetting a sheet or oiling a piece of paper. Most current translucent or rear projection screens are etched rigid plastic, etched glass or mat-surfaced flexible plastic. They, like the front screen, must scatter light as evenly as possible over the entire audience and avoid any "hot spots." A major claim for rear projection is that it can be used in a lighted room. This is only true when the image is kept very small or where very directional screens are used which in turn limit the viewing area. It is also possible to combine the projector and screen into one unit, perhaps on a large wheeled cart for moving into position when needed. Many special presentation halls have been constructed with large rear projection screens and a complex rear projection room out of sight behind it. Very effective multimedia presentations can be made with technical assistance, rehearsal and carefully planned lessons. One projection room and technical staff can serve several audiences around it provided it is carefully engineered to prevent any stray light from appearing on any of the screens.

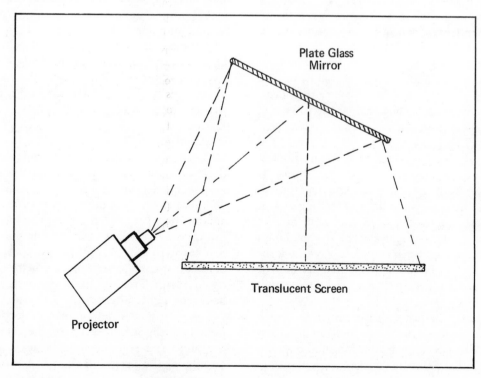

FIGURE 5.3. Rear Projection System

Cabinets with several projectors and one rear screen have been developed for use in the front of a classroom. This unit must be "programmed" before the class meets in order to have the visuals on demand.

Rear projection requires short focal length lenses, high-quality (preferably front coated) mirrors to bend and erect the images, or materials must be inserted in special ways. Overlays on overhead projectors may be impossible to use. It may be difficult to make adjustments for optimum viewing or to make changes during class time. Equally good presentations, even with combinations of materials, can be made with either front or rear projection.

Mat, beaded and lenticular surfaces cost approximately the same per square foot and translucent materials cost about twice as much.

Screens for filmstrips, lantern slides and motion pictures are normally rectangular with an aspect ratio of three by four or 1 to 1.33. This means that the screen is three units high and four units wide. (Width is the horizontal dimension.) Two by two slides, opaque and overhead projectors normally have square apertures so that square screens are desirable. Square roller screens can usually be opened to a marked position which makes them rectangular. Cinemascope or wide screen films generally require an aspect ratio of about 1 to 2.5. They are seldom used in education. This extra wide screen may, however, be used for two projector presentations in which two images need to be compared or related. Two ordinary screens could, of course, also be used side by side.

In placing a screen it is important to determine that everyone in the audience can see all of it. The bottom of the screen should be no lower than head height, or only those people in the first rows will see all the pictures. It should be noted that most captions for visual materials are placed across the bottom. The top of the screen is normally limited by the ceiling or suspended lighting fixtures. If these do not limit height, then the comfort of people in the front row may be the limiting factor.

The size of the screen is determined by the preceding height factors and also by the minimum two screen widths and maximum six screen widths viewing distances. Projector focal lengths and/or distances can be selected so that the screen will be filled.

If the screen is higher than the projector, which is the normal situation unless a booth or platform is used, then the distance to the top of the screen is greater so that more magnification will result. A picture wider at the top than the bottom is called keystoned. This defect can be corrected by tipping the top of the screen out until the surface is at right angles to the line of projection. Most screens have antikeystone devices included or available. Keystoning is much more pronounced with projectors such as overheads which are operated near the screen. If keystone correction is made for the overhead, then projection from the back of the room will have reverse keystoning, but it will be slight due to the much greater projection distance.

Screens are attached or supported in several ways. If the screen must be portable, a table top or tripod screen will probably be used. If permanently assigned to a classroom, a wall or ceiling mounted roller screen in a metal tube is most common. Large classrooms and auditoriums often are equipped with electrically operated roller screens which are very convenient but expensive. Much less expensive large screens are operated by rope and pulleys or they may be permanently stretched on a frame by means of lace (cord) and grommets (reinforced holes). Large lace and grommet screens may be suspended above an auditorium stage. Various folding frames and folding screen surfaces are available for special purposes. The important consideration is to have a good and proper size screen properly located where it is regularly needed.

In areas of high humidity it is important to have a screen that is mildew resistant. For an auditorium stage and for some other situations, flame resistance would also be important.

The poor projection associated with so many educational situations is usually due to the lack of light control. In order to make dark areas on a screen it is far better to darken the room than try to increase projector lumens or find a more reflective screen surface.

Selection

There is a greater variety available in screens than any other audiovisual equipment. A screen may be needed about a foot wide to attach permanently to the wall of an individual study carrel and another may be needed a dozen or more feet wide for remote electrical operation in an auditorium. Screens may be for front or rear projection, large or small, square or rectangular, mat or with directional characteristics, fixed or portable, and for heavy or light duty.

The choice of front or rear projection is determined by the arrangement of the room or rooms, the projection equipment to be used and the philosophy of the person in charge. A translucent rear screen is usually part of a package that has been integrated for a particular purpose rather than selected for a variety of uses. An exception is the boxlike unit containing a mirror and screen to use for individual or small group viewing with almost any projector. Shorter than normal focal length lenses are commonly used with rear screens. If a choice of rear screen surfaces is available, selection should be based on brightness and uniformity of picture, uniformity of brightness at all expected angles, absence of any "hot spots" where the direct beam from the lamp is evident, absence of reflection from room lights, absence of visible graininess from expected viewing distances and absence of a second, or ghost image.

The size of the screen for normal use is directly related to visibility factors for viewed materials in common use. Most educational visual materials are made for optimum viewing by an audience seated no closer than two screen widths and no farther than six screen widths. A screen to be viewed by an individual in a carrel at a distance of 3 feet should be between 6 and 18 inches wide. A screen 6 feet wide would be recommended for viewers from 12 to 36 feet.

The size of the screen is also related to focal lengths and projection distances. These relationships are given in the projection tables and discussion of optics. It is preferable to choose a screen slightly larger than needed rather than smaller so that no light will spill over the edges.

The size of the screen is also related to height of the ceiling in the room since the only area of any screen that is visible from the back of the room is that part between head height and light fixture or ceiling height. It is better to have a smaller image that is entirely seen than a large one that is partially obscured.

Ambient or stray light in the room may also dictate a smaller than optimum screen size in order to get enough light for visibility. A small bright picture may be preferable to a larger and "washed out" one.

Small screen sizes are seamless and large ones may be made from strips that are cemented together. The maximum size without seams may be different for different companies and different fabrics or surfaces. The seams are usually more visible in room light than they are when pictures are projected.

The shape of the screen is determined by the shape of the material to be projected. Movies and filmstrips are always rectangular, with the larger dimension horizontal. Other materials may be square or with the longer dimension vertical. Ver-

tically oriented material may be difficult to see in a classroom with very limited distance between head height and lighting fixture height. It might be remembered that a square screen can always be used as a rectangle, but not vice-versa. With roller screens of any type, wear and tear is most apt to occur at the top or bottom, and repair often means a shortened screen, so it might be wise to start with a square one for this reason.

Screen surface is primarily determined by the desired distribution of reflected light. If the most uniform distribution is wanted for all students at all viewing angles, then the mat screen is needed. Probably all testing based on projected material should be done with mat screens. In areas with high ambient light levels and limited viewing angles, one of the directional screen surfaces should be selected. It should spread light as evenly as possible over the whole audience and be free from annoying glare or dark areas.

For classroom use the wall or ceiling mounted roll-up screen and the tripod types are most commonly used. If a screen can be permanent or semipermanent, then the wall or ceiling model will be less expensive, wear much longer, be better located for viewing and be completely out of the way. The portable tripod screen can be moved and set up anywhere. Large tripod screens may be very awkward to move and difficult for ordinary people to set up.

Portable screens come in a variety of qualities from inexpensive home movie types to rugged audiovisual types. For constant school use the latter will be most economical and satisfactory. The legs on tripod screens are a common source of annoyance. Magnetic leg locks are very helpful.

For auditorium use a very large screen is needed to assure visibility from the back seats. If a loft, rope and pulleys are available, then the lace and grommet screen on a light frame is the least expensive and very satisfactory. Most theater screens are of this type with a border to mask the lace and grommets. If there is insufficient height to fly a rigid screen, it may be on the back wall or hinged for movement. Large screens on rollers may also be hand operated with a rope and pully arrangement. Screens larger than about 7 feet wide on spring rollers similar to the classroom wall or ceiling type may be difficult to operate and easily damaged. Electrically operated wall or ceiling screens come in all the larger sizes and are very satisfactory, but at relatively high cost. The operating controls can be placed in a projection booth or at any convenient location or at more than one location. The mechanism should be automatically turned off when the screen is all the way up or down.

If overhead projection is going to be regularly used, then antikeystoning devices should be ordered with the screens. Brackets are usually used with wall screens and permanently attached arms are recommended for tripod screens. The lace and grommet auditorium screen can be easily tilted by tying back the bottom. Roller auditorium screens are not easily antikeystoned.

Operation

Laboratory operation is limited to three portable tripod screens with different surfaces and other characteristics. The screens should be carefully handled since they are easily damaged. They should be kept as much as possible out of other people's way and put away as soon as the work is completed. Any available projector can be used for the light and angle comparisons. There is some danger of injury when releasing a large screen under tension. This can be minimized by holding the movable part when the release button is operated. The laboratory operation sheet is at the end of this section.

Maintenance

There is little maintenance required on fixed screens. Portable models should be checked to see that mechanical parts are tight and operating properly. A little lubricant will improve the operations of the sliding rods but may get on someone's hands. Rips in fabric are difficult to repair without special equipment and materials which are available at larger audiovisual dealers. Ordinary pressure sensitive tapes do not make good repairs for screen fabrics. Mat surfaces and some lenticular surfaces can be washed. Beaded surfaces should only be brushed with a clean soft brush or should be vacuumed. Screen cases should be checked to see that no edge is rubbing against the fabric during withdrawal or return. Only soft-tipped pointers should be permitted with glass-beaded screens.

Assignment III.   Checked_____

Name_____

Date_____

PROJECTION SCREENS

1. A. Mat screen make and model _____ price_____

    B. Beaded screen make and model _____ price_____

    C._____ screen make and model _____ price_____

2. Stand each screen up with legs at the bottom and determine the method for releasing the legs so that they will spread apart. How are the legs released?

    1A. _____

    1B. _____

    1C. _____

3. A bar or rod in the top of the case or standard must be extended upward to hold the top of the screen fabric. Describe the method for releasing the extension rod.

    1A. _____

    1B. _____

    1C. _____

4. Holding the top of the standard and extension rod, rotate the screen and roller case so that it is horizontal with the screen fabric coming out the top.

5. Lift the top of the screen fabric by the ring at the center and attach it to the top of the extension rod.

6. Release the extension rod lock (see step 3) while holding the rod and raise the rod and screen fabric to the full open position. What is the height and horizontal width of each screen? (Screen dimensions usually include any border.)

    1A. _____

    1B. _____

    1C. _____

7. It is possible to have the fully opened screen near the floor or raised up in order to have the bottom at eye level. For this operation, both the top of the screen and the roller case must move up the standard. Some roller cases have catches to release and some move automatically. What is the minimum and maximum height off the floor possible for each fully opened screen?

1A. _____ minimum _____ maximum

1B. _____ minimum _____ maximum

1C. _____ minimum _____ maximum

8. Tripod screens often have antikeystone arms for overhead projection. Indicate the maximum extension possible in inches and the number of intermediate stops.

1A. _____ 1B. _____ 1C._____

9. Overlap the three screens so that a projector can be focused on parts of the three surfaces simultaneously. View the images from the projector (zero degrees) at about 20 degrees and at about 45 degrees and compare the results. (Note that this is not a carefully controlled experiment.)

1A. 0°_____ 20° _____ 45° _____

1B. 0°_____ 20° _____ 45° _____

1C. 0°_____ 20° _____ 45° _____

10. Considering the 2W and 6W (W=screen width) limitations on screen use, what would be the recommended minimum and maximum viewing distances, in feet, for each screen.

1A. _____

1B. _____

1C. _____

11. Considering the directional characteristics determined in step 9, what special uses might each screen have?

1A. _____

1B. _____

1C. _____

12. Return each of the screen fabrics to its case by reversing the steps above. Be careful to hold the extended rod and/or fabric when the lock is released and slowly let it down and into the case. Pack up the screens in the smallest and most portable way and return them to storage.

chapter 6

Overhead Projectors

Background

An overhead projector is designed to take the image from a transparency placed on a horizontal stage and project it upward into a lens-mirror combination and then over the operator's shoulder to a nearby screen. The operator normally faces the group. A relatively short focal-length lens is used to provide a large image in the limited space at the front of a room. The transparency is usually placed directly on the open stage with no carrier or automatic position device. Figure 3.6 shows a common arrangement of the optical system.

The first overheads were designed specifically for three by four lantern slides to be used on science demonstration tables. This size also found application in reading programs, particularly remedial reading. Since no carrier or holder was used on the stage, some part of large slides or transparent materials could be used. Some reading materials about 7 by 4 inches are designed to be slid over the lantern slide size opening.

During World War II larger overhead projectors were developed for military instruction. Several different sizes and shapes of apertures were developed to go with matching transparencies. 5 by 5, 7 by 7, 8 by 10, 10 by 10 and 10-inch round apertures were developed and marketed by various manufacturers. The Vu-Graph with a 10- by 10-inch format developed by Beseler proved to be the most popular machine, and most work has continued on this size machine. Other sizes and formats are available for special applications.

The overhead projector has advantages that are rapidly making it the most common audiovisual machine. It appears that most classrooms will soon have an overhead projector as a permanent and integrated instructional tool. The overhead projector puts so much light on the screen (about 2000 lumens) that it can be used in a normally lighted classroom, so long as direct sunlight is excluded. The teacher can face the class and maintain direct eye contact with the students. Many materials can be projected, including small opaque or transparent objects, extemporaneous writing on plastic sheets or rolls, locally prepared thermal or diazo transparencies and commercially prepared transparencies. They can be pointed to, added to, progressively disclosed, overlaid, moved and exposed for any desired period of time. Complex transparencies with overlays or moving parts can be used with teacher explanation to develop difficult concepts.

The large amount of light on the screen is due to the large and efficient optical system used. The large transparency, usually 10 inches wide, needs to be magnified only a few diameters (seven, in order to fill a 6-foot screen), and a large condenser lens under the transparency collects a large amount of light from the lamp. The projection lens often has an f number of 3.5.

The traditional all-glass condensing system and three element projection system for materials of this large size resulted in a heavy and expensive, although effective machine. Several ways to make a lighter, smaller and less expensive machine that would still accept 10-inch transparencies and produce an acceptable screen image have been developed and used.

A condensing lens must be used very close to the transparency in order to converge all possible light from the lamp into the projection lens and onto the screen. This lens must have a diameter equal to the diagonal of the transparency and a short focal length in order to collect much light from the lamp in a reasonably small box. If such a lens were plano-convex and made of glass it would be several inches thick, heavy and expensive for the usual 10-inch transparency. Since the curvature of the lens surface, rather than its thickness, bends the light rays, a special lens called a fresnel lens is usually used. The fresnel lens is diagramed in Figure 3.5. In common practice two thin plastic sheets with precision optical surfaces in the form of many concentric rings are cemented together to make a light, inexpensive and effective condenser lens. The curved surfaces are inside the plastic "sandwich" in order to protect them and keep them clean. Only flat outer surfaces need to be cleaned. A cross section is shown in Figure 6.1. Fresnel lenses can vary widely in the efficiency and effectiveness with which they concentrate the light uniformly through the transparency and into the projection lens. Some of them spill undesirable light into the face of the operator and students. Observers near the screen may see the concentric rings of the lens on the screen.

A small glass condensing lens near the lamp is used on some projectors in conjunction with the fresnel lens at the stage. This lens may increase the light collected and/or distribute it more evenly.

If a projection lamp is used that must be burned base down, then a mirror must be used under the stage to direct the light upward through the stage. This mirror is ordinarily large and made of glass with the silver safely on the back, as with ordinary mirrors.

Many current projectors use a tungsten-halogen lamp that can be used in a horizontal position directly under the middle of the stage with a reflector under it and possibly a glass condenser lens immediately over it.

A few overhead projectors make use of a large concave mirror under the stage rather than condenser lenses. A machine is shown in Figure 6.2. This system spills much less light in the operator's face than fresnel lenses. This type of machine is often used for large group instruction in auditoriums.

In order to reduce weight and cost, the traditional three element projection lens is often replaced with a one or two element system that provides excellent light-gathering ability with reduced resolution. Projection lens focal lengths in overheads vary from 10 to 14 inches, which of course determines how far from the screen the machine must be placed for a certain size image. The trend seems to be toward shorter focal lengths in order to put the operator nearer the screen. The focal length of the projection lens must be matched to the condensing system.

Since overheads are used between the audience and the screen, special precautions must be taken to see that all parts of the image can be seen by all of the audience. A common arrangement, Figure 6.3, is to put the machine on a low stand so that the operator can sit beside it. The bottom of the screen should be about eye level and the top tilted out to avoid a keystoned image. The image may be projected upward toward the screen by tilting the whole machine or the projection optics, or both.

Overheads are often turned on and off a large number of times during a presentation. The lamp is on only during those times when the image is

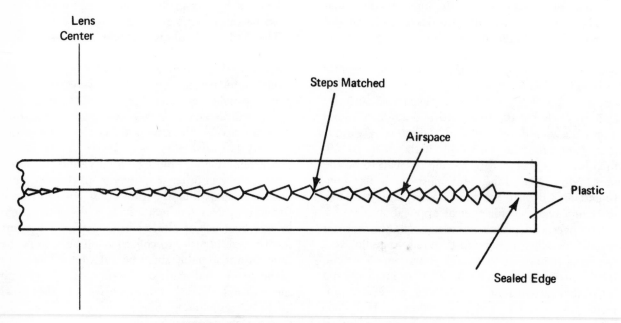

FIGURE 6.1.   Fresnel Lens for Overhead Projector

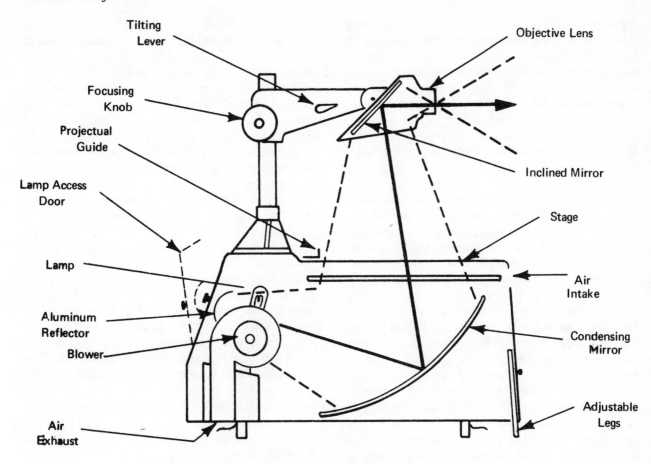

Tilting Lever

Focusing Knob

Projectual Guide

Lamp Access Door

Lamp

Aluminum Reflector

Blower

Air Exhaust

Objective Lens

Inclined Mirror

Stage

Air Intake

Condensing Mirror

Adjustable Legs

FIGURE 6.2. Overhead Projector Using Condensing Mirror

Courtesy Tecnifax Corp.

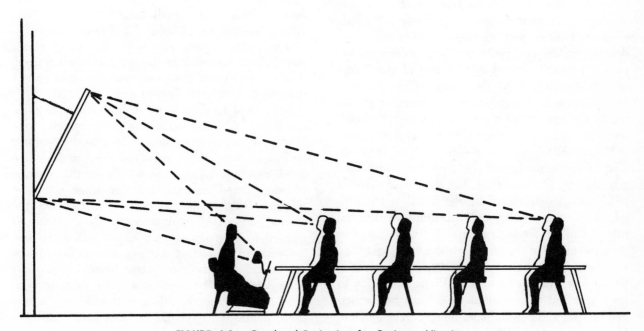

FIGURE 6.3. Overhead Projection for Optimum Viewing

actually needed and turned off when attention should be directed to the teacher. Although turning lamps on and off decreases their life, this is not an important consideration with this machine. Most projectors have a three position switch or a thermostat so that the fan can be left running after the lamp has been turned off.

Much work that was once done on chalkboards is now done on overhead projectors.

Selection

The use of the overhead projector has been growing rapidly so that many brands and models are available for a variety of applications and budgets. The safety, optics and convenience wanted must be weighed against size, weight, heat, stray light, noise and cost for the particular setting or settings in which it will be used.

For safety, the Underwriters Laboratories seal or Canadian Standards Association seal should be evident on the machine, or the individual items that make a machine safe must be checked. A three-wire self-grounding power cord should connect any exposed metal parts of the machine to the plumbing in the building. Any access doors that might be opened by an operator or student should automatically turn off the power, or no contact with live terminals should be possible. Lamp changing should be possible without hazard. Any place that might be grasped during movement of the machine should be free from sharp edges.

The optical system is designed to put a bright, sharp and evenly illuminated image of some size on a screen at some distance. Light output is measured in lumens and determined by lamp, condensing system and projection lens. Screen brightness is also determined by the area of the image and the screen surface. The sharpness or resolution of the image is determined by the number of elements and quality of the lenses. Available projectors have one, two or three elements, and sharpness ordinarily increases with the number of elements. The corners and edges of the image ordinarily have considerably less resolution than the center. Needed resolution is determined by the detail in the projected picture (chalkboard substitute to detailed maps) and the sophistication of the user. The corners and edges of the image are also less bright than the center. The center-corner ratio should generally be no more than two to one. Projectors differ in how near and how small an image can be focused and how far and how large an image can be focused. If very small or very large images are needed this

characteristic needs to be checked. Some machines have a lamp-positioning device to obtain optimum illumination and absence of colored bands when the projection distance is changed.

Convenience is made up of many features such as position of the post that supports the projection lens, position and nature of the lamp switch, thermostat, focusing device, elevating device, plastic roll position, roll length and direction of movement (if any), accessory outlet for plugging in another machine, spare lamp storage or changer, power cord length and provisions for storage, power cord "captive" or detachable, positioning device for transparencies such as pins or a backstop, and one or more shelves or trays for transparencies before or after use.

Size and weight are related to portability. If the machine is to be permanently installed for immediate and regular use, then a large and heavy machine may be desired. If one machine must be used in several locations, then low weight and small size are desirable. Suitcase models are also available. Machines should not be moved or jarred when the lamp is on. The size and height of the projection lens and mirror may interfere with visibility. On the other hand, if they are too low, they may interfere with writing and using overlays on the stage.

Heat becomes a problem when projectors are used for long periods, which is now common. The cooling system should keep the stage and any parts handled, such as the switch, cool enough for comfort even after an hour of continuous operation. Transparencies with large dark areas trap heat in the machine. Masks placed over transparencies may also increase heat to the discomfort point. Excess heat may melt or dry out the markers used. The fan and vents must be protected from papers that might block the flow of air.

Stray light or glare is not a problem with infrequent use of a machine in a brightly lighted room. It may be an important annoyance to the operator and/or students when regularly used under some conditions. Machines vary widely in this characteristic. Tinted acetate over the stage or lower wattage lamps will reduce illumination on the screen and glare simultaneously. Good design is needed to keep image very bright and glare low.

Overhead projectors are not designed to make any sound, but the fan system inevitably does. If a machine is to be used in a quiet room or with a microphone nearby, then a machine with low noise should be selected. Machines may emit more noise in some directions than others.

Maintenance

Overhead projectors are usually operated for long periods in the presence of dirt which is sucked into the machine by the fan. Heat also sets up convection currents around the machine. Chalk dust is a common problem.

The various lenses and optical components of the machine need to be cleaned according to the instructions given under lenses. The projection head assembly may need special attention. If it is not sealed it can be opened with simple tools and the parts cleaned. The first surface mirror must be cleaned with extreme care, and it is better to leave it dirty than run any risk of damaging it. Sealed optical units and three element lens barrels should be opened and cleaned only by a specialist.

Both surfaces of the stage and fresnel lens need to be cleaned often since dirt on these surfaces appears in enlarged form on the screen. A damp cloth usually works well. Glass window cleaners should not be used unless they specifically include plastics. If markers that are not water soluble have been used, it may be necessary to clean the surface with lighter fluid or something similar. Lighter fluid should be used only on a cold and disconnected machine since it is obviously flammable.

If overheads are moved or handled roughly, the optical system may get out of alignment, which will result in uneven illumination or colored areas on the screen, particularly in the corners. The reflector and lamp can be adjusted with simple tools. The projection head may need slight repositioning.

A little lubricant may be needed on the focusing and elevating devices. Fans sometimes get foreign matter around them and become noisy. A drop of oil on the fan motor shaft may also quiet it, but care must be taken to keep oil from getting on optical surfaces.

The usual nuts and bolts that loosen through use and vibration should be tightened. The power cord should be inspected for wear and to see if it gets hot, particularly at the ends, during extended use. If it does get hot, it needs a new end or complete replacement.

Overhead projectors should not be left in automobiles or any places where the sun might use the lens for a burning glass. This is most apt to happen early in the morning or late in the afternoon.

Assignment IV. Checked _____

Name_____

Date_____

OVERHEAD PROJECTOR

1. Brand and model of machine_____UL approval?_____

   price_____

2. Power requirements: volts _____ watts _____ amps _____

3. Power cord length?_____ Is power cord attached to machine? _____

4. Maximum transparency aperture of stage (H″ x W″)_____

5. Is a transparency holder or locater provided?_____

6. Projection lens focal length_____f number_____number of elements _____

7. Location of post for holding projection lens_____

8. Method of focusing lens_____ Elevating device _____

9. What is the minimum image size (horizontal dimension) that can be brought into sharp focus?

   _____

10. At what distance will this machine fill a screen 6 feet wide?_____

11. Projection lamp code_____. How changed?_____
    (Do not touch tungsten-halogen lamps with bare fingers.)

12. Convenience features: (check if included)

    A. Spare lamp storage _____   B. Plastic roll_____

    C. Marker storage_____   D. Cord storage _____

    E. Shelf for transparencies_____   F. Accessory AC outlet _____

13. Location of on-off switch _____

14. Provision for cooling machine after lamp is turned off _____

15. Make a cross section labeled diagram showing all optical elements.

chapter 7

Overhead Transparency Makers

Background*

The overhead projector requires some kind of transparent material usually called a transparency in order to go beyond the projection of small opaque objects or silhouettes.

Many transparencies are made directly on blank sheets or rolls of transparent plastic with special pencils, felt- or nylon-tipped pens, pressure sensitive tapes, dry transfer letters and symbols and so forth. These may be made extemporaneously in front of a class, or at a desk during class preparation. The time and limited talent of most teachers result in crude or infrequent transparencies.

Most of the materials that teachers wish to project on the overhead projector already exist on paper or they can be most easily and competently done on paper. A primer typewriter produces good size letters for transparencies. Several machines are available which will enable a person with little training and skill to prepare a full size (one to one) and direct reading (positive) transparency from marks on a sheet of paper without the necessity for a photographic darkroom, trays of solutions, drying and the like. Several darkroom and room light photographic process machines are available

that will do excellent work, but they are not considered here.

Many prepared paper masters for the local production of overhead transparencies are available in books or sets. Paper masters can be purchased in quantity at low cost for selected conversion to more expensive transparencies when and if needed.

The most common method for making transparencies from paper is the infrared or thermal copying process shown in Figure 7.1. There are a number of companies making compact and easily used machines. In practice, a sheet of specially prepared plastic is put over the material to be copied and exposed to intense radiant heat for a few seconds. Heat concentrates wherever the black marks are in contact with the plastic, resulting in distortion of the plastic. This is a reflex process in that heat goes through the plastic to copy the image from the first surface of the paper, which may be opaque and have an unwanted image on the reverse side. Ink on the paper results in similar distorted areas on the plastic. When it is put on the overhead, light is scattered, diffused or

*Some of the material included in this selection first appeared in a series done by the author for *The Instructor*.

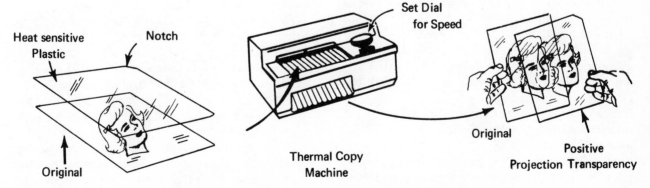

Heat sensitive Plastic   **Notch**

Original

Set Dial for Speed

Thermal Copy Machine

Original

Positive Projection Transparency

FIGURE 7.1. Thermal Copying Process for Transparencies

blocked by the distorted areas so that a dark image appears on the screen. Some plastic sheets have a transparent coating that turns black when heated. These sheets produce transparencies that are easier to read when on the overhead or when inspected without projection.

A great variety of positive and negative transparency materials are available for heat-transfer machines. Most users will concentrate on one or two that produce good results at low cost and then occasionally use colors and reversals for special effects.

The popularity of the thermal process is well deserved. It will produce usable transparencies in a few seconds at low cost from most printed material, and ordinary lead pencil and carbon copies. Materials can be handled in ordinary room light and stored indefinitely without special precautions.

The usual thermal copy process has certain limitations. Only carbon and metal-based inks will absorb enough heat to make a good image on the plastic sheet. Most colored pictures and lettering will not reproduce at all. The process is essentially high contrast with few intermediate tones. Neither fine detail nor large black areas can be faithfully copied. Since the material must roll through the machine for exposure, only single sheets, 9 inches wide, may be copied. As with all processes not involving lenses, no change in size is possible.

In order to copy images not printed with carbon-black inks and continuous tone images, a number of photographic machines and processes requiring neither lenses nor darkrooms have been developed and perfected in the past few years. All of these require a second or transfer step in order to capture a light rather than a heat exposure and develop a projectable image on plastic. A light sensitive transparent sheet is put between a light source and the material to be copied and exposed about half a minute. The light source is very bright so that relatively insensitive materials can be used and handled briefly in ordinary room light. The latent image is then transferred to a special sheet by means of pressure and heat or developing and fixing solution. Most of the machines will copy from a bound volume without injuring the material in any way. Many existing office copy machines can be used for transparency preparation simply by ordering the appropriate materials. Even though colored materials can be copied, the copies are always black and white.

The diazo system for preparing transparencies requires an original or master with opaque ink on transparent or translucent material. In practice, black ink and a variety of rub-on and stick-on materials on tracing paper are most often used. The diazo equipment provides a bright source of ultraviolet light and concentrated ammonia fumes.

Diazo film, often called foil, is coated with a faint yellow incomplete dye that is barely visible under ordinary conditions. The incomplete dye can be broken down or destroyed by exposure to ultraviolet light for a minute, or even a few seconds if sufficiently strong. The incomplete dye can also be completed into any desired colored dye by exposure to ammonia fumes for a brief time. The final color will depend on the incomplete dye materials applied to the film.

A hand prepared or printed diagram, lettering, figure and so forth on tracing paper is put over and in close contact with a sheet of diazo film and exposed to ultraviolet light. The ink is usually placed face down against the upward-facing diazo coating identified by the notch in the upper right corner in order to produce the sharpest images. The ultraviolet rays easily go through the tracing paper but not the opaque lines. The incomplete dye protected by the lines is then completed by exposure to ammonia to form a transparency that can be projected without further treatment. The simplest machines make use of a sheet of glass and a printing frame arrangement under an intense lamp in a box and then a large jar (often called a pickle jar) with concentrated liquid ammonia absorbed in a sponge at the bottom. The process is slow (several minutes), but excellent work can be done. Faster machines make use of a rotating drum that carries the tracing and film around one or more long tubular lamps. Large "white print" machines are available that will expose and process diazo films about as fast as thermal machines. Many specialized diazo materials are available for local production of a great variety of high-quality transparencies. A sepia reversal film requiring immersion in water makes negative transparencies from positive originals.

Thermal copy materials will last indefinitely in ordinary lighted rooms at ordinary temperatures. They can be purchased in quantity and stored with no precautions. Both phototransfer and diazo materials are somewhat light sensitive and have limited shelf life. They should be kept in a refrigerator or purchased in quantities that will not be kept more than a few months.

Many schools are now leasing or purchasing electrostatic copying machines, primarily for making paper copies of letters and other printed materials. They are rapid and easy to use and pro-

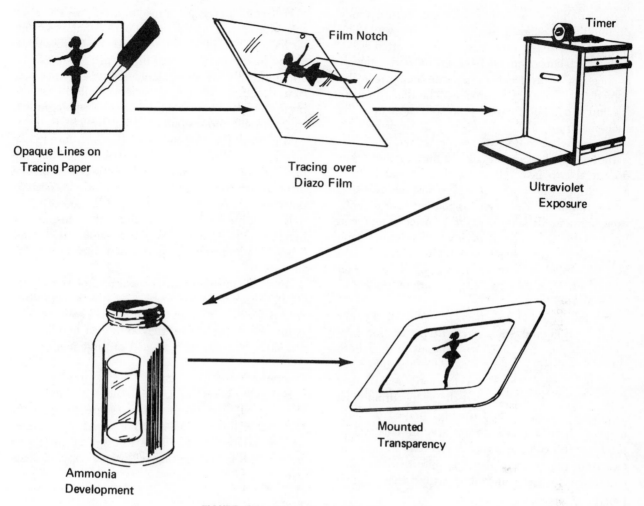

FIGURE 7.2. Diazo Process for Transparencies

duce good dry copies of anything that can be read. Transparency materials for these machines are now being developed and users of the machines should keep in touch with their suppliers concerning the availability of projection materials.

Paper copies of almost anything made on an electrostatic machine make excellent transparencies by running them through a thermal copy machine in the usual way. If thin or tracing paper is used for the electrostatic copy, it can become an intermediate master for the diazo process.

The actual ink commercially printed on paper can be transferred to a sheet of transparency material by color-lift. Only ink that has been printed on clay-coated paper (used in most magazines employing full color pictures) will work. The presence of clay can be determined by rubbing a moistened finger over any blank white area. With experience and care, full color transparencies can be prepared from most pictures in popular magazines. None of

the other processes except full color photography can do this.

A number of companies now produce machine-material combinations that make color-lift easier and better. In all cases the original paper sheet is destroyed and a wet, messy and time-consuming process must be used. However, the resulting full color transparencies seem well worth it to many teachers.

The transparencies produced from printed paper by any of these methods can be mounted for easy projection and filing.

Selection

Overhead transparency makers are selected on the basis of what they will copy and the speed or volume needed.

Black carbon or metal base ink or marks on one side of tracing paper can be copied by any of the processes easily, quickly and with good results. Most masters deliberately prepared for easy transparency production in schools are made in this way. The thermal and diazo processes are most often used.

Black carbon or metal base ink on both sides of a sheet of translucent or opaque paper can be copied only by a reflex process. The diazo process is eliminated. The thermal process is most used for such materials, although others will work as well with somewhat more expense or bother.

Individual sheets with colored inks, or those that do not absorb heat, eliminate the thermal process as well as the diazo process. Several two-step photographic transfer systems making use of both exposure and development or transfer stages can be used for this type of original. Continuous tone and high contrast black and white materials can also be copied with these machines. All copies, no matter what the original, are in various shades of black and white.

Any ink on individual sheets of clay-coated paper can be lifted and transferred to plastic for projection. Some machines will permit lifting both sides of the paper at one time. After the inks have been transferred, the paper must be soaked and then rubbed off and flushed down the drain. The exposed ink is dried and usually sprayed or laminated with clear plastic to preserve it. This system produces full color and continuous tone transparencies from paper originals. It is unfortuantely time consuming, messy, and it destroys the original paper.

If the paper to be copied is bound in a book or magazine and cannot (or should not) be removed, then a book-copying machine must be used. Most of the machines mentioned must have individual sheets that can roll through the units. Book copy machines have flat or slightly curved exposure units that come out to an edge over which the sewn or bound edge of the page, the gutter, may be placed. Since the whole page must be exposed at once it may be rather slow. If the book is tightly bound it may be very difficult to get good exposure near the binding. Electrostatic book copying machines have been highly developed to make paper copies which can then be used on a simple thermal machine for making a transparency.

If the material to be copied must be changed in size, then a machine using lenses must be used. Many electrostatic machines can enlarge or reduce copies by a moderate percentage. Photographic systems involving copy cameras and enlargers can be used to make almost any size copies from almost any size original.

Full color overhead transparencies from various size originals in books or on paper that must not be harmed are often wanted and most difficult to obtain at the present state of the art. They can be obtained only by using large sheets of full color film in large cameras at several dollars per exposure. At the present time, full color photographic work is most often done in the 2 by 2 size.

Machines for thermal and diazo transparency production should be selected somewhat on the basis of speed and volume. If many copies of one original are to be made for distribution, speed is particularly important. If local production of originals is combined with individual transparency making, then speed is of little importance since other operations can be performed during machine processing.

Another consideration is flexibility of the equipment. Teachers often want to make a composite transparency with items from several sources. Machines differ widely in their ability to handle various edited pieces in a special arrangement.

Operation

Separate sheets for transparency production by the thermal and diazo processes are included at the end of this section. If other copying machines are available, the same sheets or special ones may be used for experimenting with their copying characteristics.

Maintenance

The major maintenance operation with copying machines is cleaning. Many masters involve soft pencil, black ink that is not entirely dry, rub-on and stick-on letters that involve waxes and pressure sensitive tapes that use nondrying adhesives. A particular problem occurs when dry transfer materials made for the low heat diazo process are unwisely used in the high heat thermal process. All of these materials can accumulate on the plates, drums, cylinders and belts used in these machines. The result can be a dismaying collection of spots, or ghosts of former masters on each new transparency. The machine should be disconnected from the electrical outlet before cleaning.

Glass can be cleaned with any of the common glass cleaners. A dry or damp soft cloth will remove most materials. A dry cleaning solution (with care) may be needed to remove some adhesives that have transferred.

The belts on rotating machines should be cleaned only with a slightly damp cloth or the special solvent that is made for this purpose and supplied by the maker of the machine. A sheet of paper or plastic sometimes gets wound around a roller and must be unwound.

Complex machines should be cleaned only according to the instructions provided with the machines.

Many poor transparencies result from poor contact between the master and copy. This will result in weak or blurred images in whole or in part. Some means of adjusting the tension or pressure is provided on many machines. It should be adjusted only according to the instructions with the particular machine. Machines that have a weighted cover to hold materials in contact may need an additional push for thick, wrinkled or other problem materials.

The exposure on all these machines is rather critical. The only adjustment ordinarily needed is changing the time for exposure and processing. Most people use relatively few materials in a standard way so that optimum exposures can be marked on or with the machine in order to avoid waste. Unknown combinations can often be tried out with narrow test strips of the sensitive film. Internal adjustments of thermostats and voltage may be needed if the adjustments above do not produce good copies. These should be made by company service representatives.

Assignment V(a).   Checked _____

Name _____

Date _____

feed face up
from this edge

THERMAL TRANSPARENCY

1. Machine brand and model _____ Machine price _____ UL approval? _____

2. Electrical requirements: Volts _____ Watts _____ Amps _____

3. Film or transparency used: Code _____ Size of sheet _____ Price per sheet _____

A. Prepare a labeled diagram in this box with a soft-lead pencil.

B. Attach a small newspaper clipping to this area with a small piece of #810 Scotch tape at the X position.

X

C. Try various pens and pencils on the lines below by writing their names.

1. _____
2. _____
3. _____
4. _____
5. _____
6. _____
7. _____

D. Include samples of regular and primer typing in this box.

Obtain exposure time or setting from instructor

Transparency notch over this corner

Assignment V(b). Checked _____

Name _____

Date _____

DIAZO TRANSPARENCY

feed face up
from this edge

1. Machine brand and model _____ Machine price _____ UL approval? _____

2. Electrical requirements: Volts _____ Watts _____ Amps _____

3. Film or foil used: Code _____ Size of sheet _____ Price per sheet _____

A. Prepare a labeled diagram in this box with a soft-lead pencil or a pen and india ink.

B. Attach various rub-on and stick-on letters, symbols and tapes in this box.

C. Try various available pens, pencils and typing on the lines by writing their names.

1. _____

2. _____

3. _____

4. _____

5. _____

6. _____

7. _____

D. Prepare silhouettes cut out of opaque paper and attach them with #810 Scotch tape.

Obtain exposure time or setting from instructor

Film or foil notch under this corner

Note: A sharper transparency will result from putting the film emulsion in contact with the printing, and processing the combination upside down.

chapter 8

Opaque Projectors

Background

The opaque projector has been used for nearly as many years as the lantern slide projector, and many opaques were also referred to as magic lanterns. The early models made use of two lamps on each side inside a box with a small inverted picture on the back wall and a lens on the front. The faint picture could be projected on a screen in a totally dark room, but since no mirror was used to reverse the image sideways, the result was an erect but mirror image on the screen. Later versions of this machine used incandescent lamps with somewhat brighter pictures.

About forty years ago the Bausch and Lomb Optical Company and Spencer Lens Company (now American Optical) developed machines using mirrors and powerful incandescent lamps to produce correct reading images of sufficient intensity for darkened classroom viewing. Any projectors, including opaques, made by the first company were often designated Balopticans and the others. Delineascopes. Opaque projectors are also called reflectoscopes and the British call them episcopes. The early opaques would project a page size only up to 6 by 6 inches. Most modern machines will project a 10- by 10-inch page.

The opaque projector produces a relatively dim image (about 1/10 as bright as a lantern slide projector) due to the inefficient reflection of light from a piece of paper rather than transmission of light through a slide. A powerful lamp, usually 1000 watts, is surrounded by mirrors to concentrate and distribute evenly as much light as possible on the paper or object to be projected. The usual arrangement is shown in Figure 8.1. The light falling on the paper is scattered or diffused in all directions. Some of the light from the paper is intercepted by the large mirror at a 45 degree angle and directed into the projection lens for recreating the image on a distant screen.

It should be noted that no condensing lenses are used as in transparency projectors in order to concentrate the light from the lamp into the projection lens. It is possible to transmit far more light through a transparent slide than to reflect it from a piece of paper.

The mirror used to reflect and invert the image inside the projector must have its reflecting surface on the exposed outer surface rather than protected by glass as is the case with all ordinary mirrors. If the reflecting surface were behind the glass, then two images would appear on the screen slightly displaced. The image produced by the first surface would be about a tenth as bright but still visible. The first surface mirror must be treated with care because the reflecting surface is ordinarily silver, which can be easily scratched or tarnished.

The projection lens in an opaque projector is the most expensive type used in any common audiovisual equipment. It must have a long focal length, usually about 18 inches, in order to fill a typical classroom screen at a convenient distance, about 9 feet. Teachers often wish they could use the opaque from the back of the room, which would require a much longer focal length lens and a much larger machine to hold it. It would seem more advantageous to use a shorter focal length similar to the overhead projector (10"-14") and use both of the machines from the front of the room, but this arrangement is not commonly available.

The projection lens must have great light-gathering ability due to the relatively dim image it must use. The projection lens must also have high resolution because fine print from a page is often projected for a group to read. This means

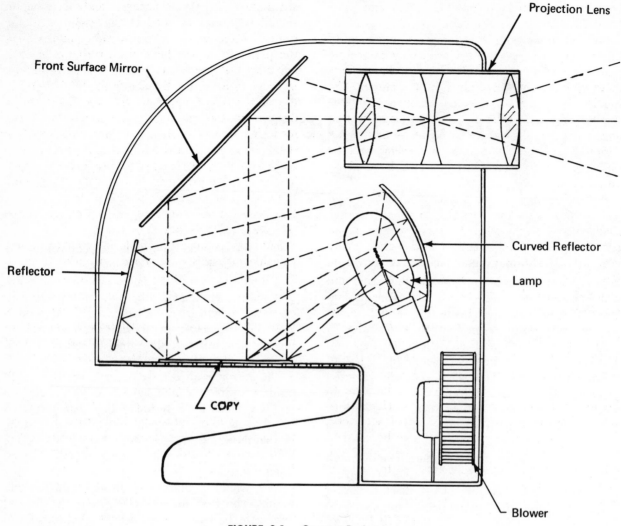

FIGURE 8.1. Opaque Projector

that a three element highly corrected lens must be employed.

An opaque projector should never be left with its projection lens uncovered in a place where the sun might enter and use it for a burning glass. This can happen in an automobile or in a classroom early in the morning or late in the afternoon.

A fan is used to keep heat under control, and it may also be used to keep loose sheets of paper anchored in position either by blowing down on the sheet or sucking air from under it. Even so, the machine may get very hot, and pages from a book or magazine may flutter while they are projected.

A pointer consisting of a small movable beam of light is often included to call attention to certain areas of the screen image.

The opaque projector has the tremendous advantage over transparency projectors of using any

flat paper printed materials that can be read. Many reasonably flat materials such as coins, leaves, insects and keys can also be projected. Colors will be accurately reproduced on the screen. Student papers in any subject or grade can be immediately projected for comment or group discussion.

The opaque also has disadvantages that make it one of the least popular audiovisual machines. It is awkward and uncomfortable to use, and it produces relatively dim images. It would be specially useful in a windowless classroom. It also has a special application out of class time for enlarging small printed materials to trace for poster, chalkboard or bulletin board use.

Selection

Selection of an opaque projector should be based on size of material projected, method of holding material, magnification, quality of screen image, heat control, optical pointer, elevation and electrical considerations.

Most opaques project material up to 10 by 10 inches. This can easily be checked by projecting part of a yardstick inserted in the machine. Some older projectors project only 6 by 6 material. The smaller the material projected to fill a standard-size screen, the greater will be the magnification and hence readability of fine print. A newspaper clipping projected by a 6 by 6 machine will be more easily read than when projected by a 10 by 10 machine.

Paper material to be projected must be held down inside of the machine, due to the air currents produced, by tempered plate glass over the paper, but this can be bothersome and the glass easily gets dirty. It also may break, and it cuts down on screen illumination. Air from the fan may be directed downward onto papers to anchor them or it may create a vacuum under them. The most difficult thing to project is a page from a bound book. Even with the plate glass system, it may be difficult to get all of the page in focus at once. If uniform size sheets are to be projected, a metal mask with a cutout of the correct size may be used to frame and hold the sheets. Several machines provide endless belts, perhaps an optional accessory, in order to permit a series of sheets to be cranked through the machine one immediately after the other. An alternative arrangement is to tape or paste sheets together into a long strip which is moved through the machine by hand.

The device used to hold the projected material is usually called a platen. It is often on a compound hinge so that various thicknesses of material can be inserted. Its opening will limit the thickness of books that may be projected. There may also be a door at the rear of the machine, and under the front surface mirror, for inserting specimens, electric meters or even a watch. This door may also be used for cleaning the machine and changing lamps.

The focal length of the lens determines the degree of magnification. Ten- by ten-inch machines usually have an 18-inch focal length lens with a diameter of about 5 inches to produce an f number of about 3.5. Longer focal lengths may be available, but they have lower f numbers and even less bright images than normal. The 18-inch lens unfortunately locates the projector neither in the front of a classroom nor in the back but rather among the students.

Opaque projectors normally have very high-quality three element lenses that produce very sharp images. The sharpness and illumination should be nearly as good at the edges as in the center. If the screen is considerably higher than the projector, it will be necessary to tilt or anti-keystone the screen as with overhead projectors in order to have the distance from the lens to the top of the screen approximately the same as the distance to the bottom. A compromise must be made between large images in order to read fine detail on much printed material and small images in order to obtain images as bright as possible. The size of the projected image is also limited by the amount of movement permitted in the projection lens. A small bright image may be impossible to focus due to limited lens travel.

Efficient cooling is desirable in order to protect valuable materials and provide operator comfort. Machines can be compared by projecting and observing high temperature thermometers. Although the heat of some machines may seem dangerous, there are few cases of actual burning, and even those come from copy actually getting in contact with the lamp. A glass heat filter is available for some machines if critical materials are to be projected for extended times.

An optical pointer is desirable for calling attention to a particular area on the screen. This may use light from the projection lamp or a separate bulb. It may project an arrow or the filament of the bulb.

Opaque projectors are so large that they may obstruct the view for many students. In order to make better visibility the machine may be put on a low stand so that all can look over it. This arrangement requires that the operator sit behind the machine. It also requires that the machine be tilted upward by 20 or more degrees, and few machines provide elevating devices that will permit this. The alternative is to prop it up on books or boxes.

Opaque projectors are usually equipped with handles and considered portable, but few teachers want to lift them. It would seem wise to purchase opaques with wheeled stands and always keep the two together.

A heavy power cord is required for this machine and it should be long enough to reach the outlet without an extension cord. Fifteen feet is usually enough. A "captive" power cord is one that is permanently attached to the machine so that it cannot be borrowed, lost or forgotten. The on-off switch should be convenient to the operator where he would normally be located. It is difficult to replace projection lamps on some machines. A clip or other device to hold a spare lamp would be desirable.

A cover may be ordered to reduce the accumulation of dirt when the machine is not in use.

Operation

The opaque projector should be placed securely on a stand or table and the power cord plugged directly into a wall outlet or a heavy-duty extension cord. A piece of white paper with sharp printing should be placed in position for projection by lowering the platen, or cranking it into position with the endless belt. The lamp should be turned on and the lens and elevation devices adjusted to recreate an image on a convenient screen or surface. It may be necessary to darken the room in order to do this. If the image is not direct reading, rearrange the paper until it is.

An assignment sheet for machine operation is included at the end of this section.

Maintenance

The opaque projector requires little maintenance because it is a relatively simple machine.

The lamp and associated mirrors for illuminating the copy should be cleaned with any convenient cloth. The lamp may be difficult to remove due to its awkward position. The usual medium prefocus lamp is removed by pushing in and then turning counterclockwise a fraction of a turn until it is free. If there is any doubt about lamp removal, the lamp code should be checked to determine the type of base and removal procedure if it is not given in the available literature.

The large front or first surface mirror between the copy and the lens needs special attention due to the possibility of damaging it. The best procedure is to keep all foreign matter off it except dust and to brush that off with a clean soft brush or simply blow it off. If foreign matter such as finger prints are on the mirror, extreme care must be exercised in removing them. If they are not numerous or severe, it probably would be better to leave them there. Lens tissue and a little moisture can be very carefully used on the surface if necessary.

The projection lens should be cleaned as described under lenses.

Adjustable and removable parts should be checked to be sure they operate and will not come off and be lost. The platen opening device may need regular lubrication for quiet and easy operation.

Fans on older models have oil holes for a few drops of electric motor or projector oil. The fan should be checked for immediate starting, proper operation and free air circulation. Pieces of paper sometimes get sucked up into the machine.

If the power cord has screw terminals at either end, or an in-line switch, the connections should be tightened.

OPAQUE PROJECTOR

1. Brand and model of machine_____ list price_____

2. Power requirements: volts_____ amps_____ watts_____

3. Does it have a cooling fan?_____Where located?_____

4. Projection lamp code_____ (Do not try to remove the lamp.)_____

5. How many mirrors are used in the machine to concentrate light on the paper?_____

6. What is the maximum size of material that can be projected?_____

7. What is the maximum thickness of a book that can be projected?_____

8. Projection lens: focal length_____ aperture_____ f number _____

9. How is the machine elevated?_____ maximum elevation?_____

10. Does it have an optical pointer?_____

11. How close to the machine can a picture be brought into focus?_____

12. At what distance will maximum size material just fill a 70″ x 70″ screen?_____

    _____

13. How is material oriented in the machine for correct reading on the screen?_____

    _____

14. Describe the arrangement for holding individual pictures or sheets so that they can be easily pro-

    jected without flutter. _____

    _____

15. How does screen brightness compare with transparency projectors?_____

    _____

16. About how high is the type on this page when projected onto a 70″ x 70″ screen?_____

    _____

17. Can objects such as a flat stone or watch be inserted and projected?_____

    _____

chapter 9

Filmstrip and Two by Two Slide Projectors

Background

The lantern slide made still projected pictures a regular part of class instruction, but the large size, cost and fragility limited their use. The first strip-film system (before 1930) consisted of a roll of positive projection film 2 1/4 inches wide that went on spools attached to a special slide carrier in lantern slide projectors. Thirty-five millimeter (1 3/8″) filmstrips were very soon developed in order to make use of nonflammable and very inexpensive theater motion picture film. Attachments for lantern slide projectors were once used to project filmstrips, but they soon gave way to special projectors for the purpose. The filmstrip projector is now one of the most common audiovisual machines.

Several names have been applied to the 35-millimeter length of film with about fifty still images on it. *Filmslide, slidefilm, picturol, still film, strip film* and *filmstrip* have been used. Filmstrip is the most common term today and will be used in this discussion.

The filmstrip has become very popular due to its low cost, very easy operation and wide variety of excellent materials. An often cited disadvantage is the predetermined order of the pictures.

Early filmstrips came in single or double frame formats and the distinction is shown in Figure 9.1. It should be noted that single frame filmstrip images are oriented with the horizontal dimension of the picture across the strip and double frame filmstrip images are oriented with the horizontal dimension of the picture along the strip. This means that machines for projecting both formats must have a means for revolving the strip holding mechanism 90 degrees. The double frame image has twice the area and a strip must be twice as long for the same number of frames. Since the double frame image does not need so much magnification, it may be

brighter and sharper on the screen. Double frame filmstrips can also be made directly in typical 35-millimeter cameras. Single frame filmstrips can only be made in special 35-millimeter cameras called half-frame models. Single frame filmstrips are not

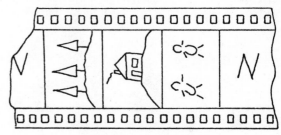

Single Frame Strip

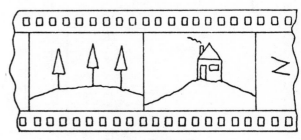

Double Frame Strip

FIGURE 9.1. Single and Double Frame Filmstrips

usually made into 2 by 2 slides because of their smaller image size. Practically all filmstrips encountered today are single frame format with a vertical dimension of .668 inches and a horizontal dimension of .885 inches. A width of 9/10 inch is usually used in rough calculations.

Filmstrips are ordinarily stored in cans about 1 1/2 inches both in diameter and height. Special strips longer than a hundred frames require larger cans. Filmstrips can be safely stored without any special air conditioning except in very moist areas.

Filmstrips are ordinarily projected with a simple straight-through optical system as explained in the section on the basic optical system. To create a classroom screen image 6 feet wide from a filmstrip image requires a magnification of eighty-seven diameters which requires a precise and high-quality optical system. Filmstrips are also used for individual study with low-power projectors, or even small viewers that use a small lamp and magnifier rather than a projection system. Filmstrips are also used in large group or auditorium instruction. Current projectors use from a 100- to a 1000-watt lamp. Lamps of more than 200 watts must have a cooling fan with an attendant noise problem. A typical filmstrip projector is shown in Figure 9.2.

Single frame (ordinary) filmstrips are fed into the top of the projector with the images upside down and backward as viewed from the back of the projector looking toward the screen. Correct projection can be assured simply by holding the filmstrip by its edges (to avoid finger marks and

scratches) and manipulating it until a correct reading of the title is obtained by direct viewing toward a convenient light. Then the strip is rotated exactly 180 degrees (inverted) and fed into the machine. Some form of holder for the filmstrip above the machine is usually provided.

During projection the filmstrip is usually held flat and in precise position by pieces of glass in front of and behind the film. This arrangement of glass pressure plates prevents the strip from suddenly moving out of focus as it expands with the concentrated heat and light from the lamp. Since the pressure plates can cause the film to stick or become scratched, they must be opened slightly during film advance or they must have raised edges or tracks so that friction occurs only along the nonprojected edges. Filmstrip emulsions absorb moisture from the air in high humidity locations and sticking between the plates may occur. Special pressure plates have been developed to overcome this difficulty or the filmstrip can be dried out over the projector lamp exhaust or otherwise. Some

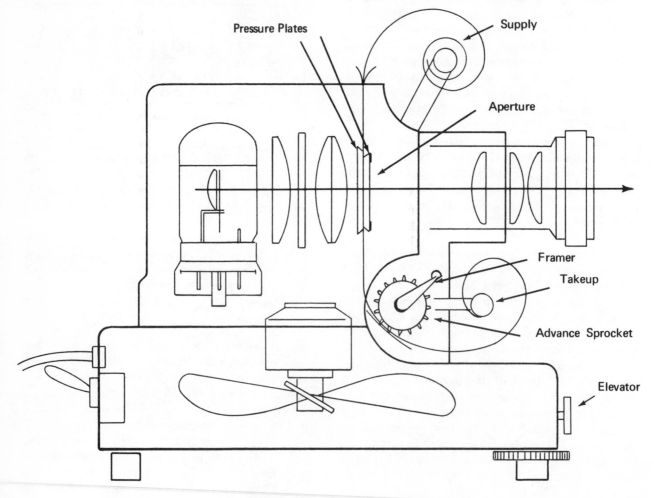

FIGURE 9.2. Filmstrip Projector

projectors do not have one or both of the glass pressure plates and rely on holding the frame at the edges and possibly curving it somewhat to control the change in focus during warm-up.

The filmstrip is accurately positioned in the aperture for projection of exactly one whole frame and advanced exactly to the next frame by a toothed sprocket wheel under the aperture, which engages several of the sprocket holes of the filmstrip. Older machines had a hinge and latch so that the outer glass pressure plate, sprocket guide and lens housing could be swung open for insertion and positioning of the strip. "Push-in" threading is now almost universal. This requires a filmstrip with a trimmed and smooth end which is pushed into the top opening as the advance knob or wheel on the sprocket is rotated. Some advance wheels rotate clockwise, and some counterclockwise. If no arrow is included on the knob or housing nearby, trial and error is required.

In order to advance the filmstrip one whole frame at a time, a positioning cam or other device is attached to the back of the sprocket wheel.

Since there are four sprocket holes per frame or image, there is only one chance in four that one whole picture will appear on the screen until the machine has been "framed" with a device called the "framer." Framing is usually accomplished by pushing in and rotating the advance knob in either direction until one whole picture appears on the screen, or a small lever beside the knob accomplishes the same purpose. Once adjusted, framing should be automatic for the rest of the strip.

Some machines, particularly for small group or individual use, feature sprocketless construction. With these machines the filmstrip is pushed in as with other machines, framed simply by pushing the strip up or down, and advanced by pushing down on a lever which brings out one or more teeth to engage and advance the film one frame. The strip may be rapidly advanced or withdrawn by hand so long as the advance lever is not in use.

The simplest machines have no filmstrip take-up mechanism and a coil of film appears under, inside or in front of them during operation. Afterward the film is removed by hand, handled only by the edges, and put back in its container. Some machines have take-up or rewind devices to dispose of the projected film or even put it back in the can ready for storage or reuse.

Some filmstrip machines make use of cartridges that are inserted in the top of the machine and automatic threading occurs when the advance mechanism is operated. Unfortunately there is no compatibility between cartridges of different manufacturers.

Projectors that will also project double frame filmstrips must have a second and double size aperture mask that is put into position for the larger image. A common arrangement is to remove or swing aside the single frame mask and expose a permanent double frame mask. Another arrangement involves sliding upper and lower sections of the mask to change its size. The whole filmstrip holding mechanism must also be revolved ninety degrees in order to project double frame strips. As pointed out before, double frame strips are uncommon at the present time.

Remote advance mechanisms are included with or available as accessories on some filmstrip projectors. Most of these are solenoid or electrically operated plunger types that will advance the strip one frame at a time. Both forward and reverse operation requires a more complex device, but with obvious advantages. Remote on-off and focusing controls are desirable but seldom available. The usual remotely operated machine must be carefully adjusted before operating it at a distance.

Automatic advance mechanisms are built into some filmstrip projectors that are used in connection with a record player that is attached. Special thirty and fifty hertz tones (inaudible with the reproducers used) recorded on the record hold and then advance the filmstrip at the desired time. Many sound filmstrips for this combination machine have been produced, but usually they have been used for sales and industry training rather than in schools. Oftentimes the sound is recorded on one side with the inaudible signals and on the other side with an audible tone to signal when an ordinary machine should be advanced. Some automatic machines have a cord and control for remote operation as well. Instructions for synchronizing the sound and pictures usually appear on the record or on the instruction sheet.

The 2 by 2 slide became popular late in the 1930s in order to project color slides made with 35-millimeter cameras. This miniature camera and color film processed commercially has made the 2 by 2 slide tremendously popular for home entertainment. Teachers have been much slower to make their own slides for teaching purposes. Many commercial slides for education are now coming on the market.

Two by two slides have an advantage over filmstrips in that any number of pictures can be arranged in any order for a particular audience and purpose.

Other size slides between 2 by 2 and lantern slides have been developed. Square slides 2 1/4 and 2 3/4 inches are sometimes encountered in connection with intermediate size cameras, but they have never become popular in or out of schools. Adapters for lantern slide projectors are available, or special projectors may be purchased.

Two by two slides made from standard 35-millimeter cameras have the same image dimensions as a double frame filmstrip, 23 by 34 millimeters or 9/10 inch by 1 3/10 inches. Other miniature cameras produce somewhat larger image sizes that can be mounted within the 2 by 2 outside dimensions of the standard slide. Two popular sizes are 30 by 30 millimeters often called instamatic and 38 by 38 millimeters often called super slides. The three image sizes within the 2 by 2 frames are shown in Figure 9.3. Older projectors and some newer ones may have difficulty in illuminating all the corners of the larger image 2 by 2 slides.

Slides are normally mounted in cardboard frames by commercial processors. Various combinations of metal, glass, plastic and tape have been used for special mounting of these slides. Glass covering both sides of the film will protect it, make cleaning easier and prevent the change in focus caused by heat and expansion. Glass slides have disadvantages of extra thickness which may interfere with some slide changers, and if the interior becomes cloudy, cleaning is difficult. Newton's rings, which appear on the screen as intricate patterns of faint colored bands, may also be produced by interference in the thin layers between the glass and film.

Various manual slide changers have been developed to make it easy and foolproof to project a rapid succession of 2 by 2 slides. The slides in most machines must be inserted upside down and backward as with lantern slides and filmstrips. Much help is provided by a thumbmark in the lower left corner when viewed correctly without projection. It goes in the upper right corner when standing behind the machine and facing the screen. With the usual slide carrier alternate slides are put in the opening on different sides of the machine and the viewers see parts of two slides during changing.

Semiautomatic slide changers generally require that all slides go in the same side or top and an opaque window cuts off the light as the slides are changed. Some experimentation may be needed if no directions come with the machine. A special release or device is often necessary in order to retrieve the last slide.

Automatic and remote control slide projectors have become very popular for home entertainment and they are gaining rapidly in education. Most of these machines require that the slides be inserted in a special tray or container that is matched to the machine. Unfortunately, several different incompatible systems are used. Some machines will project a stack of slides that are not in any tray. Many hours can be wasted moving slides from one tray system to another. These machines can usually be remotely focused, advanced and moved backward through the sequence established when the slides were inserted. Random access to the slides is provided only with a complex device that costs much more than the projector to which it is attached. Most machines do not provide remote on-off control, so someone near the machine is often asked to do it. A built-in timer on automatic machines will automatically advance the mechanism at one or more predetermined intervals. Automatic focus-

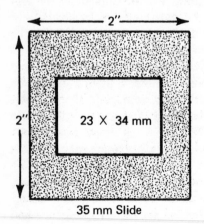

35 mm Slide

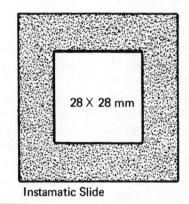

Instamatic Slide

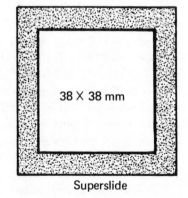

Superslide

FIGURE 9.3. Two by Two Slides

ing is incorporated in some machines in order to compensate for the change in focus that occurs when the slide heats up in the projection beam and expands to a new position. (This phenomenon is called popping.) Many projectors use a much simpler but less effective preheating system to reduce the tendency of the slides to go out of focus as they are projected.

Manually operated combination filmstrip and 2 by 2 slide projectors have long been popular since the same optical system easily projects both of them. If both single and double frame strips are also included, then the machine may be called a tripurpose projector. A simple operation is required in order to remove or swing away the filmstrip handling mechanism and insert the slide carrier. This machine is an exception to the usual caution against purchasing combination or multipurpose machines.

Selection

A wide variety of filmstrip and 2 by 2 slide projectors is available for a wide variety of educational settings. The characteristics of the machine should fit a particular educational need.

The first consideration is the material to be projected. Machines are available that project only single frame filmstrips, both single and double frame strips, only 2 by 2 slides, or any combination of the three. As the machine is designed to handle more than one kind of material, it becomes more complex, more expensive and more difficult to set up and operate properly. Larger numbers of single purpose machines should be weighed against smaller numbers of multipurpose machines at the same cost.

There are several convenience items that should be considered in connection with the setting where the machine will most often be used. The power cord should reach the electrical outlet without an extension cord. The machine should elevate sufficiently to reach the screen in optimum position without the need for props. It should be possible to level the machine so that the image is square with the screen. If remote operation is included the cord should be long enough to reach the operating position and include the desired functions. If portability is needed then all cords and a spare lamp should store in the projector or a case that is provided with the machine.

Various amounts of light, measured in lumens, are needed for various settings from an individual viewing a 12-inch wide picture in a carrel to a thousand students viewing a 12-foot wide picture in an auditorium. Lumens on the screen are primarily determined by the lamp wattage and lens f number, but the efficiency of the lamp and condensing system are also important. Lamp watts vary from 100 to 1200 watts. Most projection lenses have an f number of 3.5, but f 2.8 lenses are often available at extra cost. The f 2.8 lens (assuming optimum design of the system) will provide the same increase in illumination that would be obtained by going from 750 to 1000 watts. The faster lens would of course cost more, but this must be balanced against the greater cost of lamps, power and increased heat necessary to obtain the same screen illumination. Projectors should be tried in the setting where they will most often be used, and only that amount of illumination should be purchased that is needed for optimum viewing.

The center to corner brightness ratio on the screen should not be distracting even with the largest materials that will be projected. A ratio of 2 to 1 is often specified as the maximum allowed.

Various focal lengths of lenses are available to obtain different degrees of magnification. Four-inch and five-inch focal lengths are most common for classroom projection. Two- and three-inch lenses are often used for individual and small group use, and 7- and 9-inch lenses are often used for auditorium use. When special short or long focal lengths are considered, the f number must also be considered and the cost may rise rapidly. It may be better to compromise on the projector and/or screen location rather than obtaining an unusual lens. A special auditorium lens alone will often cost as much as a standard projector.

Zoom or variable focal length lenses are becoming common in order to adjust the size of the projected image to the size of the screen with a minimum amount of projector and/or screen moving. The zoom lens also permits an operator to mix various sizes of slides and quickly adjust the projected image size to fit the screen, provided he has carefully located the projector to cover the ranges of materials and focal lengths. Zoom lenses have f numbers that mean the same as those on fixed focal length lenses. The focal length is given as a range such as 4 to 6 inches. The focal length of a zoom lens may also be given in millimeters, and one inch equals 25.4 millimeters (mm). Zoom lenses should be checked for the sharpness or resolution of the projected image. Zoom lenses that will provide a wide range of focal lengths, with a low f number and high resolution, are much more expensive than fixed focal length lenses.

Projectors with lamps above 200 watts are usually cooled with fans in order to keep the slide temperature below about 212 degrees Fahrenheit even with long-term operation. Some fans make a very objectionable amount of noise and the exhaust may be uncomfortable for the operator or a nearby student. This can easily be checked in the location where it will be used.

The constancy of focus as the slide or filmstrip heats up in the projection position should be considered. As the piece of film rapidly heats up in the center it expands while the edges remain constant. The expanded film usually pops outward on the emulsion side, and the screen image goes out of focus in the center. Various ways of minimizing this phenomenon are used, and the results can be checked on the screen.

If filmstrips are to be used, it should be easy to insert and advance the film, and once framed each image should be accurately positioned on the screen even though both forward and reverse are used.

If thick glass slides are to be used, special checking is needed because some machines handle them much better than others. This is particularly true with remote and automatic models.

Operation

The operation of a filmstrip and/or 2 by 2 slide projector is included on the assignment sheet at the end of this section. The machine should be completely removed from its box or case during operation for adequate cooling.

Maintenance

The condensing and projection optical systems of these projectors, except for the filmstrip gate, are very similar to those discussed in the basic optical system and lantern slide projectors. All surfaces that need cleaning should be accessible without tools. No adjustments are ordinarily needed.

The filmstrip pressure plates and gate need regular and special attention. As pointed out before, the filmstrip frame must be held in position during projection and this is usually done with two glass pressure plates held together with springs. Any dirt on the filmstrip is usually wiped off on the plates, and any particles that lie within the projected aperture will apear much enlarged on the screen. Dirt on condenser and projection lenses cuts down on illumination but is seldom visible on the screen. Another concern is that foreign matter in this pressure plate area may scratch every filmstrip through its entire length.

In most filmstrip machines it is possible to remove the whole pressure plate assembly without tools for careful inspection and cleaning. These are optical surfaces and should be treated like lenses. A toothpick or plastic scraper may be needed to remove some materials from the plates. The raised edges on many plates that are designed to ride on the edges of the strip and protect the picture area may become rough or otherwise interfere with free motion. If this happens new ones should be ordered for replacement.

After thorough cleaning and reassembly the working mechanism should be checked for smooth operation, perfect screen images and undamaged film.

The framing, advance and slide changing mechanisms may need a very small amount of lubricant such as vaseline or oil for easy operation. It is important that no oil get on optical surfaces.

Fans sometimes suck foreign matter, particularly pieces of paper, up into the machine which may make it noisy and interfere with cooling. A drop of oil at the bearings will often make a fan quieter.

Vibration from the motor and handling may loosen parts which should be checked and tightened before they interfere with operation or get lost.

The mechanism of remote and automatic control systems is too complex to be included in this maintenance program and should be handled by an audiovisual technician when the need arises.

Slide changing mechanisms occasionally "jamb" which means that a slide sticks somewhere in the change cycle. This usually means that a slide that is too large, too thick or too warped is at fault. The reverse, manual or reject mechanism will usually remove the slide. If not, then the instruction manual should be consulted before any further work is attempted. Practically all this trouble can be avoided by using only standard slides in good condition.

If several people use a machine, then the spare lamp should be checked to be sure it is unused or operable. Too many people hide burned out lamps in the carton that supposedly holds a new one!

Assignment VII. Checked _____

Name _____

Date _____

FILMSTRIP AND/OR 2 BY 2 SLIDE PROJECTOR

1. Brand and model of machine _____ list price _____

2. Power requirements: volts _____ watts _____ amps _____

3. UL or CSA approval? _____ power cord length? _____

4. Material projected: single frame filmstrip _____ double frame _____

    2 x 2 slide _____

5. Projection lamp: code _____ watts _____ fan cooled? _____

6. Projection lens: focal length _____ f number _____

7. How is the front of machine elevated? _____

8. What is the maximum elevation? _____

9. Can the machine be leveled? _____ how? _____

10. Describe the method for inserting a filmstrip _____

11. Describe the method for framing a filmstrip _____

    _____

12. What happens to the filmstrip after passing through the gate? _____

    _____

13. At what distance will a single frame filmstrip fill a 45" x 60" screen? _____

    _____

14. If double frame filmstrips can be projected, describe the conversion _____

    _____

15. Describe the method for inserting 2 x 2 slides in the projection aperture _____

    _____

    _____

16. On the two slides below, write your name in the center, and place a thumb mark on each:

For hand viewing   For projection

17. At what distance will a standard slide (.9" x 1.3" aperture) match a 60" x 60" screen?_____

18. Will the machine project super slides (1.5" x 1.5" aperture) without noticeable corner cutting or dimming?_____

19. Does the machine have provision for remote operation?_____

    A. Length of cord _____

    B. Remote advance?_____ remote reverse?_____

    C. Remote on-off? _____ remote focus?_____

20. Does machine have automatic advance?_____

    A. Timer?_____ what intervals?_____

    B. Inaudible signal from tape or disc?_____

chapter 10

Two by Two Slide Makers

Background

Making photographic projection transparencies was until recently difficult and expensive. They required critical work with a cumbersome camera and several time-consuming steps in a well-equipped darkroom. Only a few teachers illustrated their class presentations with pictures they had taken.

The present availability of versatile miniature cameras, full color roll films, commercial processing directly into 2 by 2 projection transparencies and 2 by 2 slide projectors in most schools make it easy and economical for any teacher to include large and full color pictures on the classroom screen. Practically every subject and grade taught is a reflection of the world, its people and publications. Each of these can be captured on 2 by 2 slides for easy storage and projection at the appropriate class time.

Two by two projection complements overhead projection. It is very easy to prepare diagrams, maps, charts, key words and the like on ordinary size paper and convert them to overhead projection transparencies. It is presently very difficult to make continuous tone or full color transparencies for the overhead. On the other hand, full color photographs are easy with the 2 by 2 size, and hand construction is very difficult.

Two by two slides are exposed in any of a variety of miniature cameras and processed into mounted transparencies by laboratories scattered over the country. Surprisingly good results can be obtained with very little instruction or practice.

A camera contains a roll of light sensitive film that is accurately positioned in a light-tight box behind a convex lens or lenses and a shutter that permits light to enter for a very brief and con-

trolled time. The elementary type of camera is simply aimed at the subject to be photographed with the help of a view finder and the shutter or exposure button is pressed.

Most film for slides is now color film, although black and white film can be obtained and specially processed in a darkroom to produce transparencies for projection.

Close-up photography of small objects and sections of publications can be easily done with the help of a copy stand, lights and supplementary lenses.

Two by two slides have three common film sizes that are exposed in 35-millimeter, magazine (cartridge or instamatic) or other small roll film cameras. The sizes of the film within the standard mount are illustrated in Figure 3.20.

During exposure in the camera, light sensitive materials on the film are changed according to the amount and color of the light admitted. The image is described as latent because it cannot be seen until it is chemically processed in a photographic darkroom.

In order to obtain optimum exposure of the film in the camera several parameters must be considered.

Films vary greatly in their sensitivity to light, and this characteristic is indicated by the exposure index or ASA (American Standards Association) number. Very slow films have ASA ratings around ten. Most commonly used films have ratings around fifty, and fast films have ratings over a hundred. Several other film speed systems were in common use a few years ago and many exposure meters for them are still in use. Modern imported cameras often use the DIN film speed index with the following approximate equivalents:

ASA	DIN
12	12
25	15
50	18
100	21
200	24

Conversion sheets for other systems are available at photographic stores. As the sensitivity of the film increases it is possible to expose film properly with less available light, which is often of great importance. However, high-speed films usually have more apparent "graininess" (pebble effect when greatly magnified on the screen), and the color rendition is less faithful. Simple cameras may not be able to limit the light sufficiently for high-speed films under very bright conditions.

The sensitivity of film also varies with the kind of light. Two ASA film ratings are usually provided for daylight and tungsten, or ordinary artificial light. The tungsten rating is normally about two thirds of the daylight rating. Since tungsten illumination has more of the red-orange-yellow end of the visible spectrum and less of the green-blue-violet area normally found in daylight, outdoor film exposed under tungsten illumination will appear excessively red-yellow. To overcome this, special films are available for tungsten, or a filter may be used. If tungsten film is to be used outdoors, another filter is needed to avoid excessively green-violet transparencies. Filters have a "filter factor" such as 2x which means that the exposure time must be increased two times or the diaphragm must be opened one f stop to make up for the light absorbed by the filter. Fluorescent lamps and high-intensity mercury lights (now becoming common for gymnasiums, stadiums, exhibit halls, etc.) are more like daylight than tungsten. Special filters are available for these lights, but they are not ordinarily needed.

Camera shutters vary greatly in the times provided for exposing the film. Simple cameras provide only one exposure time of about one thirtieth of a second. As flexibility and cost increase much slower and much faster times are included. A range of 1/15, 1/30, 1/60, 1/125 and 1/250 second is most often used. The most versatile shutters also include 1, 1/2, 1/8, 1/500 and 1/1000 second exposures. All except the simplest shutters provide a B position which permits the exposure button to keep the lens open as long as it is actuated for low light exposure. A T position on some shutters stands for time, and it opens the shutter on the first push and closes it any later time with a second push. Both bulb and time exposures must be made with the camera on a tripod or other firm support, and a flexible cable release to operate the shutter without moving the camera is also very helpful. Fast shutter speeds are used to limit the light for fast film and high illumination situations and to prevent moving objects from appearing blurred due to their changing position during the exposure. Slow shutter speeds are needed for exposures under low light conditions.

Camera lenses vary greatly in their light-gathering ability which is usually referred to as the speed of the lens. Simple cameras have f numbers around 11 which permit exposures only under ordinary daylight conditions or with flashbulbs nearby. A better lens with an f 5.6 rating would provide the same light in one fourth the time. Most high-quality miniature cameras come with f 3.5 or f 2.8 lenses. Very fast f 2 lenses are available for extreme conditions. As the f number decreases the diameter and cost of the lens increase very rapidly. All low f number lenses are equipped with diaphragms that permit reducing the aperture or diameter so that several higher f values can be selected at will. Some common f numbers and corresponding shutter speeds for equal exposures are as follows:

f number	shutter speed
22	1 sec.
16	1/2
11	1/4
8	1/8
5.6	1/15
4	1/30
2.8	1/60
2	1/125

The f numbers and shutter speeds on various cameras may not have these exact designations. Low f numbers are desirable in order to take pictures under poor light conditions or to permit high shutter speeds to stop motion. Low f numbers have two disadvantages (other than high cost) which should be taken into consideration. The lower the f number the less the depth of field or distance between nearest and farthest object that will be in focus. A typical high-quality camera with a 50mm lens will provide the following depths of field at the indicated f number settings:

Distance setting	f2.8	f5.6	f11	f22
∞ (infinity)	64'-∞	32'-∞	16'-∞	8'-∞
20'	15'-30'	12'-61'	9'-∞	6'-∞
10'	9'-12'	8'-15'	6'-28'	5'-∞
5'	56"-64"	52"-70"	47"-78"	39"-137"

A general rule is to use the largest f number setting that can be used with existing light, film speed and a shutter speed that will not result in evidence of movement. The other disadvantage of low f number lenses is the loss in resolution or sharpness when used at maximum aperture. This defect would not ordinarily be noticed with high-quality lenses.

Some means for determining ambient light on the subject is needed since the eye is a very poor judge of light levels. Separate hand-held meters were once common, but the trend is toward building them into the camera. The meter must be set for the speed of the film and aimed at the subject; then it will provide the various combinations of shutter speed and f numbers that will give optimum exposure. Care must be taken to read the light on the subject and not the bright sky light. A new and desirable development (at added cost) is to put the light meter behind the lens so it actually measures the light that will be used to expose the film.

If the sun or artificial lights do not provide enough illumination to expose the film, flashbulbs are easily used. Most modern cameras have a built-in flashbulb socket, or a "shoe" for attaching an external unit. A battery must be included in the camera or external unit to ignite the bulb, and contacts must be included in the shutter mechanism. A short electrical cord may be needed between the shutter and flash unit. Most of these flash units position the bulb very close to the lens, which results in "flat" lighting and no visible shadows. It also tends to make reflections back into the lens from windows, glossy surfaces and people's eyes. An extension bracket or handle and a longer cord can be used to overcome these difficulties, but the result is less portability.

Several automatic cameras are now on the market that provide some combination of cartridge (no film threading and rewinding) automatic film speed setting, motorized film advance, automatic adjustment of lens aperture (f number) to go with set shutter speeds, indication of needed flash, automatic rotation of flash cubes and so forth.

Selection

Hundreds of miniature cameras for making 2 by 2 slides are available from a few dollars to several hundred dollars. As the complexity and quality of the camera increases, it is possible to take pictures that have greater sharpness (resolution), at various distances because of a focusing mechanism, with different degrees of magnification due to different lens focal lengths, at different speeds or exposure times, under a variety of ambient light conditions, with various artificial lights, with exposure indicators and/or automatic exposure controls, with delayed and automatic exposure, with various filters and with rapid film advance. Each of these characteristics will be considered in order.

The sharpness or resolution of a camera lens is determined by the number of elements and the care used in making and adjusting them. The simplest camera lenses have only one element, and all pictures will be soft or apparently slightly out of focus. Single element lenses are widely used for family pictures to be shown on small screens to relatives. They are not recommended for serious educational work. Doublet, or two element lenses are much better than one, but four or more elements are even better, and within financial reach due to the small size required to cover such small areas of film. One important reason for the popularity of small cameras is the quality of lenses that can be obtained at moderate cost.

Every lens produces its sharpest picture when it is properly focused. A lens with a focal length of 2 inches, which is most common for 35-millimeter cameras, would be placed exactly 2 inches from the film for photographing a distant scene. In order to focus on a nearby scene the lens should be moved out away from the film. The depth of field indicates the nearest and farthest objects that will appear to be in focus. The depth of field increases as the f number of the lens increases. Inexpensive fixed focus cameras with one or two element lenses have relatively high f numbers and they are arbitrarily set for acceptable focus from about 7 feet to infinity. Better lenses have a smooth and accurate means for moving the lens outward from this infinity focus position to a point where objects about 3 feet away can be sharply focused at low f numbers when the depth of field will be only a few inches. Focusing may seem to be a nuisance, but it is essential for making sharp pictures at varying distances. Focusing may be accomplished by measuring or judging the distance and setting the focusing ring on the lens accordingly, or an optical range finder may be included for determining the distance and setting the lens simultaneously. The latter system is usually preferred.

Most miniature cameras come with a standard 50mm lens. Better cameras will permit the substitution of short focal length lenses for wide angle shots or long focal length lenses for narrow angle or telephoto shots. Some cameras are so constructed that lenses can be changed without exposing a partially used film. Long telephoto lenses (over 150mm) should be used with a tripod or some means for holding the camera steady. Close-up lenses are available for attachment to the standard lens in order to copy materials on a copy stand. An adaptor ring may be needed. An alternative to the supplementary lenses is an extension tube or bellows to move the lens farther out. Supplementary lenses are often marked from plus 1 to plus 10 diopters with increasing numbers indicating increased magnification. A plus 10 lens would focus on an area about 2 by 3 inches at a distance of about 4 inches.

Enough shutter speeds should be selected to meet normal conditions. Lower shutter speeds than 1/30 second or faster speeds than 1 1/250 second are seldom needed unless pictures are to be taken under very low light conditions without flash, or of very rapidly moving objects such as football players.

A low enough f number lens should be selected to gather light under the most severe conditions expected. Lenses below f 2.8 go up rapidly in cost and they are seldom needed since fast films and flashbulbs are usually available. A high shutter speed, such as 1/1000 second, and a fast lens, such as f 2, would still be recommended for photographing sporting events, high-speed machines and so forth.

A flashbulb or cube device should be included with most cameras, and it should be connected to the shutter for automatic operation. Blue bulbs provide the proper color balance for color film. Instructions for camera adjustment come with the camera and every box of bulbs. Flashbulbs have a guide number for each film speed. With the shutter set at 1/30 second, the guide number is divided by the camera-to-subject distance in feet to obtain the proper f setting.

A built-in exposure meter is much more convenient than a separate hand-held model, although many professionals still rely on separate instruments. The exposure meter is adjusted for the speed of the film in the camera, and then it measures the light and indicates several combinations of shutter speed and f numbers that may be used for proper exposure. Many new meters are of the cadmium sulphide type, abbreviated CdS. These meters are more sensitive to light, but a small replaceable battery must be included. Behind the lens exposure meters are very desirable in measuring the light actually going to the film.

Automatic exposure control, in which the light meter actually adjusts the diaphragm (f number) in response to light, speeds up picture-taking greatly. Some of these devices measure the light over the whole picture area, and others, called spot meters, emphasize the light in the center of the image.

A timing device on some cameras permits the operator to put the camera on a tripod or other support, adjust it, then get into the picture himself. The camera will automatically expose the film after a few seconds delay.

Filters are available for many special effects, such as emphasizing clouds, minimizing haze and using tungsten color film outdoors. These filters must match the lenses to which they will be attached. Most teacher-made slides are made without any filters.

The view finder on the camera is necessary to aim the camera accurately and determine exactly what will be included. All inexpensive cameras have a separate little lens system, usually in one corner of the camera, designed to do this. For distant scenes, it does very well. For nearby objects it is often inaccurate because it is at some distance from the taking lens, and it is difficult to determine the actual area being photographed. The problem is called parallax. Both of these difficulties are overcome with single lens reflex (SLR) camera in which the view finder operates through the picture-taking lens. A complex arrangement involving a movable mirror, a pentaprism and an eyepiece directly over and behind the lens is required. The diaphragm must also be completely open during viewing and closed down to the selected f stop during exposure. No single lens reflex cameras are made in the United States. The pentaprism makes a characteristic bulge over the lens, which often identifies these cameras.

If closeup photography is wanted, the single lens reflex camera is particularly helpful for very accurate focusing and for determining the exact area being included. A copy stand for holding the camera and lights is needed. Supplementary lenses for close-up work are available for most SLR cameras. Some inexpensive fixed focus cameras have special close-up stands available that include a supplementary lens and accurate focus on a clearly outlined area without using the view finder. Flash cubes provide the needed light. A close-up photography stand is shown in Figure 10.1.

FIGURE 10.1. Closeup Photography Stand

Courtesy Eastman Kodak Co.

The film in most miniature cameras is now advanced with a single stroke lever to the next film exposure position. Others require rotating a knob which is a little slower and less convenient. A few cameras have spring- or battery-operated film advance mechanisms.

In selecting a camera to prepare 2 by 2 slides for educational purposes, picture quality features should be separated from convenience features. No good pictures can be taken without a good lens, accurately focused and accurately timed. There are many older cameras available with excellent optics at low prices because they do not have the modern convenience features. Many new cameras have convenience features and very limited optical systems.

Operation

The object of this laboratory exercise is to expose one or more outside (possibly through the window) pictures and one or more close-up pictures to be made into 2 by 2 slides by a commercial processing laboratory. The assignment sheets are at the end of this section.

Maintenance

The only maintenance suggested for cameras is outside cleaning and replacement of batteries, if they are included.

Cameras should be kept as clean as possible. They should be purchased with accessory cases and kept inside them except for actual use. A soft clean brush can be used for removing most loose dirt and dust. Lens tissue, specially made for the purpose, should be the only material used to clean the lens after loose dirt has been brushed off. All high-quality camera lenses are coated to reduce reflections, and the coating usually appears blue or purple. It may not appear completely uniform in color. In no case should the lens be cleaned vigorously enough to remove any of the coating. The viewfinder system also needs to be cleaned with the brush and lens tissue.

Batteries used for exposure meters last a long time, probably a year. Some cameras have a test device or procedure included in the instruction booklet. Photographic and audiovisual dealers have a tester for determining battery condition. Batteries for flashbulbs and for film advance are normally replaced more often. The need for replacement is indicated by erratic or nonflashing and sluggish film advance.

Cameras should not be opened beyond that necessary for film and battery replacement.

Assignment VIII. Checked _____

Name_____

Date _____

2 BY 2 SLIDEMAKING

I. Outdoor photography

1. Brand and model of camera _____ list price _____

2. Lens focal length _____ f number _____

3. Focusing range _____ (minimum and maximum distances) _____

4. Shutter speeds provided _____

5. Is a depth of field scale provided?_____

6. Type of view-finder provided_____

7. Type and location of exposure meter provided _____

8. What provision is made for using flashbulbs?_____

9. What film brand and code used?_____

10. Daylight ASA exposure index of this film?_____

11. Describe scene selected for photograph _____

12. What focus distance is best for this scene?_____

13. What shutter speed and f stop is selected?_____

14. What other combinations of speed and f stop would provide the same exposure?_____

_____

15. What combination of flashbulb, shutter speed and f number should be used with this camera and

    film for a night exposure at 10 feet?_____

_____

16. What automatic or convenience features are provided?_____

17. Exposure number(s) used for this assignment_____
    (for later identification of slide)

II. Close-up photography

    1. Brand and model of camera _____ list price _____

    2. Film brand and code_____ASA rating _____

    3. Make and model of close-up stand_____

    4. Make a labeled diagram of the camera, stand and lights in the space below.

    5. What supplementary lens is used?_____

    6. What f stop, focus and shutter speed is suggested in the instructions for this setup and film?___

    _____

    7. Exact size of area to be photographed _____

    8. Exact distance between lens and area photographed _____

    9. What is the minimum area and distance on which this system can be focused?_____

    _____

   10. Exposure number(s) used for this assignment_____
       (for later identification)

chapter 11

Public Address Systems

Background

The basic public address (PA) system consists of a microphone, amplifier and loudspeaker designed to accept sound from one place and deliver an amplified version of it to a large area as shown in Figure 11.1. It may consist of only these three elements, or it may be very complex. It may have several inputs to accommodate additional microphones, a record player, radio tuner, tape recorder, telephone lines and so forth, and several outputs for multiple speakers, earphones, recorders, telephone lines and induction loops. This discussion will be limited to the three basic components and how they interact.

A microphone is a transducer that changes sound energy into electrical energy. Since the sound is a vibratory or wave motion, the electricity produced is alternating current and of very low power. The electricity from the microphone is carried through a shielded or coaxial cable to an amplifier which produces an adjustable high-power facsimile of the input wave form. The electricity from the amplifier is carried by ordinary wires to a loudspeaker which converts electricity into sound. In the usual arrangement, a small sound in front of the microphone becomes a loud sound in front of the loudspeaker.

When a microphone sensitive to sound is placed near a loudspeaker producing its amplified sound, there is very apt to be an unpleasant result called acoustic feedback. The British call it "howl round." A small disturbance in front of the microphone immediately produces a louder version of it in front of the loudspeaker which goes back into the microphone to be amplified again. If the microphone or speaker is not moved away or the amplification reduced by turning down the volume control, oscillation, generally a high-pitched squeal, will soon

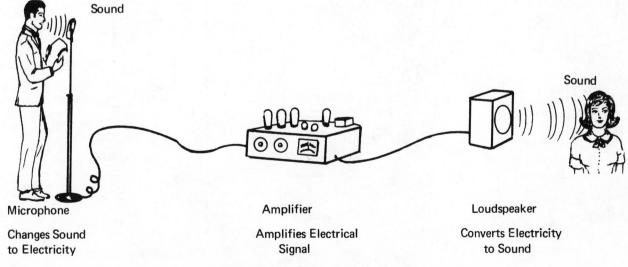

Sound

Sound

Microphone	Amplifier	Loudspeaker
Changes Sound to Electricity	Amplifies Electrical Signal	Converts Electricity to Sound

FIGURE 11.1. Public Address System

build up to the full power capability of the system. There are several ways of attacking this problem and they will be discussed later. The simplest solution is to keep the volume control just below the point at which ringing or feedback begins to occur.

There are three common methods of converting sound waves into electricity and three types of microphone result. They are diagramed in Figure 11.2. Carbon and condenser microphones are seldom used in audiovisual work and are not considered here.

The crystal or piezoelectric microphone is the simplest and least expensive type and used widely in portable audiovisual equipment. The crystal is usually rochelle salt which will produce electricity directly when vibrated. The sound is picked up by a diaphragm and transmitted by a stiff wire to the crystal which has foil cemented to two opposite faces. The crystal produces a strong and reasonable quality signal which can be fed directly to an amplifier without transformers. Crystals are fragile and sensitive to temperature and humidity. A temperature of 120° F, such as is found in a closed auto in summer or a tube-type amplifier case, will dehydrate and ruin the crystal. Ceramic microphones work on the same principle as crystal microphones but they are not damaged by heat. On the other hand, ceramic microphones produce less electricity for a given sound so the amplification must be greater. Cable lengths for these micro-

phones are generally limited to about 25 feet due to noise pickup and signal loss.

The dynamic microphone has long been a favorite for high-quality public address, radio and recording. A diaphragm picks up the sound waves and moves a coil of wire attached to it in a magnetic field produced by a permanent magnet. This interaction (generator principle) produces electricity which can be fed into an amplifier. In practice a small transformer is usually included in the microphone case to produce a low impedance (around 200 ohms) to go to professional amplifiers which contain another transformer before amplification. Such low impedance systems can use long microphone cables without appreciable loss or noise pickup. Some dynamic microphones also have a high impedance transformer winding to feed amplifiers used for crystal units. Dynamic microphones can be very rugged, reliable and high in fidelity, but at a price.

The ribbon or velocity microphone has no diaphragm. Sound entering the microphone strikes an extremely thin and lightweight ribbon stretched in a strong magnetic field. The ribbon responds to every vibration and produces an extremely faithful electrical reproduction of the sound. A built-in transformer produces low and/or high impedance outputs. Although this microphone has highest fidelity it has several disadvantages. It cannot be used outdoors because the slightest breeze pro-

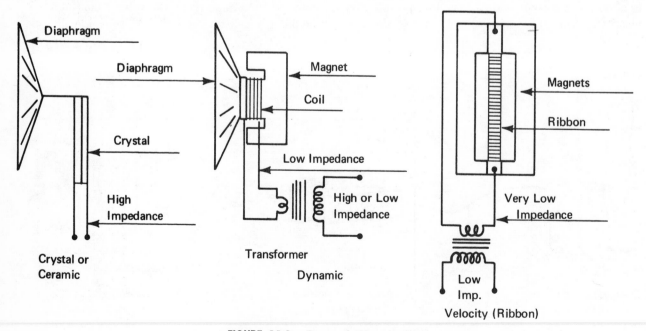

FIGURE 11.2. Types of Microphones

duces sounds like a gale. A cover called a wind screen will help. It has low output and must be used with high-gain and high-quality amplifiers. Close talking is not permitted or explosive sounds, particularly "p", will make an unpleasant burst. It also has a severe directional pattern so that a participant at right angles to the ribbon will produce much more output than a person parallel to it. This directional pattern may also be an advantage in other situations. The condenser microphone is now replacing many velocity microphones for highest quality work.

Microphones are available with three standard pickup patterns so that sound can be collected equally from all directions (omnidirectional or nondirectional), equally from only the front and back (bidirectional or figure 8 pattern) or only from one direction (unidirectional). These are diagramed in Figure 11.3.

All inexpensive microphones are nondirectional to sounds originating more than a few inches away.

Sounds will be picked up equally well whether the front or back or side is facing them. Most audio-visual microphones are of this type because they are so easy to use, as well as inexpensive.

The velocity or ribbon microphone is bidirectional because the ribbon must have sound hit its broad side rather than its edge. Those who speak must stay within limits on either side. This microphone would not be suitable for a group standing around it. On the other hand, it can be used to discriminate against unwanted sound sources such as a fan or machine. It can also have its two "dead" sides pointed toward loudspeakers on either side and one of its sensitive sides pointed toward the person speaking to provide considerable sound without feedback.

Unidirectional microphones pick up sound from only one direction. Looking down from the top the pickup pattern often looks heart shaped and is called cardioid. A high fidelity cardioid microphone is expensive but very effective in controlling feedback and eliminating various unwanted sounds. It must be carefully positioned and not moved by an inexperienced speaker. Extremely directional microphones that often look like guns are used for special public address purposes.

The fidelity of microphones may be indicated by a stated frequency response, but this must be accompanied by the decibel variation such as −5 dB. A better system is to graph the output versus frequency as in Figure 11.4.

The electrical output of microphones for a standard sound input is given in negative decibels. High output types have about −50 dB ratings and low output higher fidelity types have about −56 dB

Omni-Directional
(Non-Directional)   Bi-Directional
(Figure Eight)   Uni-Directional
(Cardiod)

FIGURE 11.3. Microphone Pickup Patterns

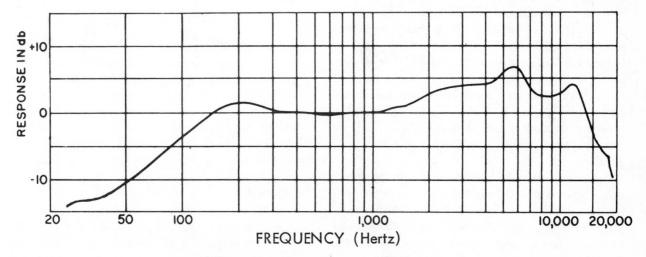

FIGURE 11.4. Microphone Output Versus Frequency

ratings. The lower output from the microphone requires higher gain from the amplifier in order to produce the needed signal for the loudspeaker.

Most audiovisual microphones have a socket for a table or floor stand with a standard 5/8 inch diameter twenty-seven threads per inch tube. Broadcast microphones often have a larger socket, and adaptors are available. Small microphones are often provided with a cord to be worn around the neck, and they are called lavaliers. Small microphones may also have a small clip-on desk stand instead of a threaded socket.

Microphones must have a connector called a jack on the end of the cable for connection to the amplifier. A connector may or may not be supplied with the microphone. Unfortunately, many types are in use that are not interchangeable. A different connector may be attached with screw terminals or soldered, or a short adaptor cord or device with unlike connectors may be used. Connectors are also designated male and female depending on whether the conducting terminal or terminals are protruding or recessed. Low impedance microphones use two wires within a flexible shield and high impedance units use one wire within a shield.

Wireless microphones have a small radio transmitter and battery included. They transmit the sound on a carrier wave to a receiver connected to the amplifier. They, of course, permit a person to move around without a tether while he is speaking. They are complex, expensive, and there is a possibility of interference from a passing police cruiser or other nearby transmitter. The battery also needs to be checked before every use.

Amplifiers take very small electrical signals from microphones, phonograph pickups, tuners, and amplify, mix and adjust them so they can be fed at usable power levels to loudspeakers, earphones, recorders and the like.

Amplifiers make use of vacuum tubes or transistors and associated electronic circuits. Vacuum tubes have been used for many years. Transistors have advantages of small size, low heat, low-power consumption and instant starting. Either tubes or transistors can provide high fidelity sound at comparable costs.

Amplifiers have one or more inputs to insert signal sources into the system. The simplest amplifiers have only one channel. A multichannel amplifier permits the selection, amplification and mixing of several sources into a composite output. The most common example is a record with accompanying commentary. A single channel may have more than one jack connected to it in parallel or a switch to select from several inputs. This multi-input system is less flexible than several channels.

An input has a jack which must match the connector to be inserted, or an adaptor must be used. (Screw terminals are used for some semipermanent installations.) An input also has an impedance characteristic usually designated high or low. Most audiovisual systems are high impedance (50,000 ohms and up), so that crystal microphones, crystal phonograph cartridges and tuners will connect directly through short-shielded cables. Higher quality systems are usually low impedance (around 200 ohms), and special transformers or preamplifiers near or inside the amplifier are required. Much longer cables can be used with low impedance systems and other advantages result. Most permanent public address and central sound systems use low impedance inputs.

Inputs may be designated high or low gain to indicate whether they are to be used with a microphone that has so little signal that high gain is needed, or with a phonograph pickup that has so much signal that only low gain is needed to drive the amplifier. Proper gain is provided when the volume control is operated near midrange for normal output. If a low output microphone (such as −56 dB) is used with a low gain amplifier, it will be necessary to turn the volume control so high that noise may be evident. The overall gain (volume control at maximum) of amplifier channels is given in decibels. Common microphone channels range from 115 to 130 dB. The decibel is a logarithmic power ratio defined by the formula

$$dB = 10 \ \log \frac{P}{p}$$

in which P indicates the power output and p indicates the power input. An amplifier that would produce 10-watts output from .0001-watts input would have a gain of fifty dB.

The gain of an amplifier is controlled with a volume control for each channel and probably a master volume control if several channels are available. Remote volume control may be available as an accessory. The output power of an amplifier is measured in watts and is related to the sound level that can be obtained with the loudspeaker. There are many ways of measuring and reporting the watts output for an amplifier, particularly for home high fidelity systems, and much confusion results. For audiovisual purposes, continuous sine wave power output, over a specified frequency range, with specified dB deviation and specified total harmonic distortion, should ordinarily be used. A portable public address amplifier for speech amplification in a small auditorium should have about 10-watts output from 100 − 6000 Hz ± 3 dB, with no more than 5 per cent total harmonic distortions (THD). One watt is sufficient for most classrooms, and 30 watts is often recommended for large school auditoriums. Outdoor systems may require 100 watts or more. If music is reproduced then a frequency response from 50 − 12,000 Hz ± 3 dB and 5 per cent THD would be recommended. High fidelity amplifiers with much more power, wider frequency range and much lower distortion are available for critical listening under ideal conditions. Amplifiers ordinarily have much greater fidelity than the microphones or loudspeakers used with them.

Tone controls are usually included in public address amplifiers in order to accentuate or decrease certain frequency ranges. The simplest tone control progressively decreases the high frequency response in order to remove record hiss, final s sound on words or to make music appear to have more bass. It may also decrease speech intelligibility. The best system is to have separate bass and treble controls that will decrease or increase both the low and high frequencies by about 15 dB as shown in Figure 11.5. Careful use of such controls can result in more intelligible speech, deemphasis of unwanted sounds, more pleasing music and some control of feedback.

Switched tone controls may be used in addition to or instead of continuously variable controls. A rumble filter cuts off low frequencies. A hiss filter cuts off high frequencies. A speech filter gradually cuts off low frequencies to make the sound less "boomy." Various other switch designations may be used such as "crisp" and "mellow" and "anti-feedback."

A volume level meter, often called a VU meter, may be included with a control to adjust it to read a certain level for a certain desired sound. It is particularly important when sound is controlled at a remote location, backstage or in a booth.

A limiter or compressor may be included in special PA amplifiers in order to maintain a relatively constant output with widely varying inputs. It is expensive to make such devices with wide control and high fidelity.

The amplifier has one or more outputs to drive loudspeakers, earphones or some other load. Jacks or screw terminals may be used. In order to transfer power efficiently from the amplifier to the load, it is necessary to have the impedance of the amplifier match or equal the impedance of the load. In order to accomplish this most PA amplifiers have an impedance selector which may be a switch, separate labeled jacks, a pigtail lead to be attached to one of a series of labeled screw

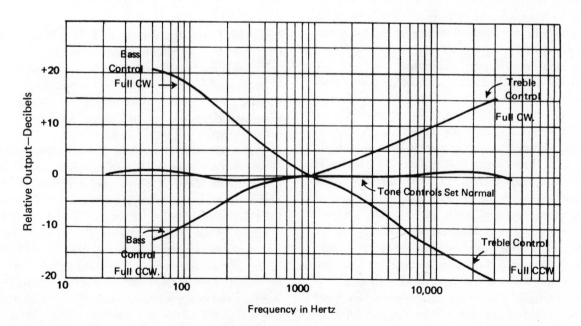

FIGURE 11.5. Dual Tone Controls

terminals or a multicontact jack to which the wires may be soldered in various ways. The most common output impedances are 4, 8 and 16 ohms. Special high impedances labeled 500 ohms, 25 volts and 70 volts may be available for connection to multiple speakers over long lines. If the proper 4-, 8- or 16-ohm tap is not used, no harm will be done to the amplifier or loudspeaker, but both maximum volume and fidelity will suffer. An amplifier should not be operated without a load such as a speaker connected to its output terminals or with shorted output.

A high impedance, very low-power output is often provided in order to feed a tape recorder input during public address. If this jack is not provided a tape recorder can be connected directly across the speaker terminals or line with small clips. It may be necessary to reverse the clips for best results.

If a hum is heard or if a slight shock is felt when touching any part of the system, the power plug should be removed and reinserted after a 180 degree reversal. Stubborn cases may require a grounding wire from the metal case or chassis (often a screw terminal labeled G) to a water pipe or the metal pipe that encloses the wiring. Grounding is particularly important when AC-operated equipment is used outdoors or near moisture of any kind. If no ground is available it may be necessary to drive a pipe into the ground.

The hum and noise of an amplifier may be reported in dB below rated output. The greater the number, the lower the unwanted sound. A 60 dB signal to noise ratio would mean that the noise was one-millionth the signal.

Loudspeakers convert electrical energy into sound. The usual construction of a cone loudspeaker is shown in Figure 11.6. A coil of wire called a voice coil is accurately suspended in a strong magnetic field provided by a permanent magnet. When alternating current from the amplifire passes through the coil, it alternately moves in or out and carries the paper cone or diaphragm to which it is cemented. The cone is attached at its outer and inner edges by very flexible materials to permit faithful response to the currents in the voice coil. Public address loudspeaker cones may move in or out a half inch or more to produce loud sounds. A 4-ohm loudspeaker has a few turns of heavy wire in its voice coil, while a 16-ohm model would have many turns of smaller wire. Either can be connected to similar impedance am-

plifier outputs with equal results unless long wires are needed, in which case the higher impedance would permit smaller diameter or higher gauge wires.

If short lengths of wire are used between an amplifier and its loudspeaker, then most any wire

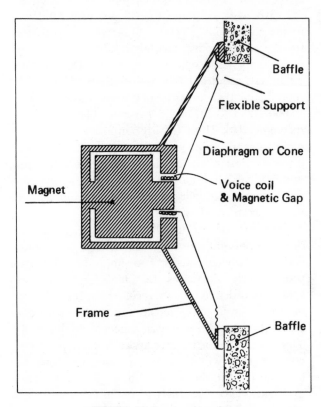

FIGURE 11.6. Loudspeaker

will suffice since the voltage and current are small. One watt into an 8-ohm speaker would require 2 volts and one-fourth ampere. On the other hand, if considerable power must be carried over some distance, then more power may be wasted in the wire than delivered to the speaker. Ordinary number 18 lamp or zip cord has 6.5 ohms per 1000 feet so that a 500-foot double length would waste about as much power as an 8-ohm speaker would use. A length of 250 feet of parallel number 22 wire (often used for speakers) would have the same impedance as a 4-ohm speaker.

Ordinary cone loudspeakers are only about 2 per cent efficient which means that 98 per cent of the power delivered is converted to heat instead of sound. To improve efficiency it is necessary to increase the magnetism, decrease the gap in which the voice coil moves and improve the coupling between the cone and atmosphere. The magnets used in loudspeakers are often reported according to the weight of the magnet rather than the strength of the magnetic field. When different magnet weights are available, the heavier one will usually provide more sound for a particular input. The gap is made as small as possible while avoiding any rubbing of the voice coil. The slightest rubbing of coil against magnet produces a very annoying scraping sound. Sometimes cones can be recentered, but they are more often replaced. In order to move more air, particularly at low frequencies, the diameter of the cone is increased. Public address loudspeakers are usually 8, 10 or 12 inches in diameter. The larger speakers do not vibrate well at high frequencies so a combination of a large low frequency speaker called a woofer and a small high frequency speaker called a tweeter is often employed for high fidelity. A crossover network is attached to the system to divide the frequencies for these two speakers.

Cone speakers must be used in boxes called baffles in order to prevent the pressure wave in front of the speaker from moving around the rim and joining the corresponding low pressure area behind. Ordinarily the greater the area and density of the baffle, the more effective it will be, particularly at low frequencies. In order to further improve the low frequency (music) response of speakers, they are often put in sealed boxes with an auxiliary opening or port in front. A common type is called bass reflex. They are not common in public address due to size, weight and the inability to store anything in the box. Cone speakers are also placed in a vertical row inside a baffle to form a sound column which produces a broad horizontal and narrow vertical sound dispersion. Column speakers are becoming common for either side of an auditorium stage.

When two speakers are used within a dozen feet of each other on the same amplifier, it is important to have them in phase which means that the two cones are moving in and out together, rather than one moving in while the other moves out. If there is any doubt about phasing, the two speakers can be put side by side and the wires from one temporarily switched while another person listens for a very noticeable improvement, particularly on music with low frequencies. (Do not short circuit the speaker wires by touching them together.)

Speakers normally are made with 4-, 8- or 16-ohm voice coils, and they should be connected to similar amplifier impedances. If two similar loudspeakers are connected in parallel, the resulting impedance for the amplifier is one half the impedance of either one. If two similar loudspeakers are connected in series, the resulting impedance for the amplifier is twice the impedance of either one. Parallel connections are usually recommended so that the signal will not need to go through one to get to the other.

Loudspeakers have a power-handling rating in watts. Ordinarily the total loudspeaker capacity should be as great or greater than the power output of the amplifier. A common 12-inch cone loudspeaker will handle about 15 watts.

Speakers may have baffle mounted volume controls made especially for the purpose. In order to maintain impedance matching while varying volume, two element controls called L pads or three element controls called T pads are used.

In order to obtain high sound levels, particularly outdoors, trumpets or horns are often used. The long horn may be folded and called a reentrant horn. The sound reproducer part is called a driver. Such loudspeakers may be 40 per cent efficient, weatherproof and practically unidirectional to beam the sound just where it is wanted. They often eliminate feedback problems by their directional characteristics. Unless they are very large, they do not reproduce bass notes well and are not recommended for music. They may be combined with bass reflex speakers for penetrating speech and pleasant music.

The positioning of loudspeakers is important for good sound coverage, particularly for speech intelligibility and treble response. Putting the loudspeakers where everyone can see them is a good rule. This usually means high and in front of the audience. Moving them off to the side will reduce interaction with the microphone but may result in distraction as the sound and person are separated.

It is important to have a trained and reliable person constantly at the amplifier controls during an important public address so that the sound will be adequate and pleasant. Unattended public address systems can assault the ears.

Selection

There is a tremendous variety of microphones, amplifiers, loudspeakers and accessories from which a public address system for a particular purpose could be assembled. For most educational purposes a portable or semiportable packaged unit with selections from suggested options would be recommended.

The first choice generally is concerned with power output and versatility of the amplifier. It is wise to err on the high side in both respects if the budget will permit, since a more powerful and versatile amplifier can always be adjusted downward for simple jobs and expanded as needed. Basic amplifiers of 10-, 30- and 50-watts output are most common. Two microphone and two high-level channels will take care of most needs. If extreme portability is needed, battery operated amplifiers are available.

Several microphone options are often available. Choosing a high-quality microphone with flat frequency response will tend to reduce feedback problems and make pleasant sound. Highly directional microphones are recommended if trained personnel will make use of their characteristics. If long lines are needed then low impedance models and proper transformers should be used. Small microphones can be hung around the neck or made unobtrusive, if that is important.

The most portable loudspeakers are built into the two halves of the case that holds the whole system. Eight-, ten- and twelve-inch models are most often used. They often come with 25 feet of wire so that they can be placed either side of the microphone. For extended outdoor use, high level or high fidelity music response, special loudspeakers need to be selected.

Accessories will make a more versatile system. A floor stand or desk stand for the microphone will often place it in a better position. A boom stand may be necessary to place the microphone over a lectern. Special clamps or screw-on microphone holders may also be used. Additional microphones may be needed for a panel or to pass around. A phonograph top is often ordered for the amplifier and internally wired to the proper controls.

Maintenance

The major maintenance problem with public address systems is trouble with the wires and connectors for power, microphones and loudspeakers. These wires are commonly on the floor where they are walked on, stumbled over, pulled on, closed in doors and so forth. They need constant checking in order to avoid lack of sound or extraneous noise, particularly when something is touched or moved.

Most connectors can be opened with a small screwdriver. Some require a small hex wrench or special tool. Most troubles with wiring occur where the bare wires are attached to the terminals or where the insulated wires come into the shell of the connector. Sometimes loose wires can be tightened with a screwdriver. More often they need to be resoldered with a small soldering iron and a spool of rosin core solder. If old, the wires and terminal may need scraping, and a tiny amount of nonacid solder paste called flux may be needed. If wires are broken inside or at the entrance to the connector, the whole cord must be cut off and the wires and shield, if included, prepared and connected as before. Connector wiring is not particularly difficult but a nuisance. Spare cords should be purchased or prepared for regularly used portable public address systems.

If a phonograph is included, it needs the same maintenance suggested for record transcription players.

Microphones should not be opened beyond the cord connector, which may need attention if it becomes intermittent when moved.

If the amplifier uses tubes, they should occasionally be pushed down in their sockets with the power turned off.

If a fuse is used, a spare in good condition should be kept with the machine. If the amplifier is totally inoperative, the replacement fuse should be tried. A circuit breaker with reset button may be used instead of a fuse.

Noise in public address systems may come from unused channels with advanced volume controls. All unused channels should have volume controls set to minimum, usually full counterclockwise.

Tone controls should be adjusted properly. Poor sound may be due to a poorly adjusted control.

Loudspeakers vibrate during use which may loosen the mounting hardware. All nuts and bolts around the speakers should be tight to avoid unpleasant rattles. The magnets of loudspeakers may magnetize the hairspring of a nearby watch and make it erratic. Any jeweler has a coil for demagnetizing watches.

Assignment IX. Checked _____

Name _____

Date _____

PUBLIC ADDRESS SYSTEM

1. Brand and model of system _____ list price_____

   UL approval? _____

2. Power requirements: volts_____ watts_____

3. Power output for loudspeakers: watts_____ freq. resp._____

   ± dB _____ THD _____ %

4. Channels provided: microphone _____ phono _____

   others _____

5. Tone control(s) provided: _____

6. Microphone number 1: make and model _____ impedance_____

   type _____ directional _____

   Microphone number 2: make and model _____ impedance_____

   type _____ directional _____

7. Phonorecord player (if provided): speeds_____ needle(s)_____

8. How is the tone arm locked for transport? _____

9. Is a tape recorder output provided?_____

10. What output impedances are provided?_____

11. How is output impedance changed?_____

12. Loudspeaker(s) provided: number _____ diameter_____

    make & model _____ impedance _____ approx. length

    of cord _____

13. Baffle(s) provided: material _____ thickness_____ size_____

14. Maximum volume control setting for "ringing" when microphone is one foot from front of loudspeaker

    _____

15. Maximum volume control setting for "ringing" when microphone is as far to side of loudspeaker as

    cord permits _____

16. Explain in your own words why the system squeals when the volume is turned up too far _____

    _____

chapter 12

Record-Transciption Players

Background

Record players have been used in homes since the turn of the century in order to reproduce speech and music for entertainment. Their use in schools for other than music courses is relatively recent.

The earliest recorders took sound from a performer, concentrated it with a megaphone and applied it to a small diaphragm connected to a needle which either embossed or cut a groove of varying depth, depending on the vibrations, on or in a wax cylinder which was rotated under it. Both Thomas A. Edison and Alexander Graham Bell invented cylinder phonographs about 1880. For playback, a needle connected to a diaphragm and horn traced the grooves and reproduced the sound. The wax was very fragile and the reproducing needle soon wore out the groove.

At the turn of the century Emil Berliner devised a means of recording on a disc similar to the ones used today. The wax disc, with varying grooves, produced entirely by the energy of the sound waves picked up by a megaphone, was electroplated and a rugged metal copy accurately produced. This metal master could be used to stamp out hundreds of copies from a hot plastic material that solidified during cooling into a durable record. The Victor Talking Machine Company became a leader in making disc records. Cylinder records competed for a time, but they could not be mass produced by the stamping process. The earliest machines were hand operated, and later spring drives and accurate governors were developed. Electric motors and electric amplification were applied to record playing about 1930.

Seventy-eight revolutions per minute was the standard disc-recording speed for fifty years. It apparently just happened to be the speed of an early machine that was successful, and others adopted it in order to be compatible. (This lesson has been lost many times in recent years.)

About 1930 the record playing industry produced major advances with electronic amplification borrowed from radio during both recording and playback. Sound no longer was required to do the work of moving the recording needle as it cut the groove in the master, and the needle in the record no longer had to do the work of moving a diaphragm to vibrate the air and recreate the sound. Much more faithful sound resulted, and the record which was losing rapidly to radio soon became a competitor.

Although embossing or indenting the groove was used on some early recorders, and is still used on some dictating machines, cutting the groove with a very sharp stylus and removing a ribbon of soft plastic material was found to produce much better results. The wax used in the early days was replaced by nitrate plastics that often produced very flammable masses of ribbon or "chip." The plastic records could be carefully played a few times or plated to make metal masters for stamping copies.

Some manufacturers preferred cutting the modulation or vibrations in the groove up and down (sometimes called vertical or hill and dale) and some preferred constant depth and lateral modulation. The former permitted putting the grooves closer together, but the latter permitted higher fidelity. Between 1940 and 1960 practically all records used lateral modulation. Stereo records use both lateral and vertical modulation (at a special angle) in order to record two separate signals in one groove.

It is assumed today that all records start at the outside and play toward the center. Inside-out records were once very common in order to use the sharpest cutting needle and the sharpest playback needle toward the center of the disc where the information is most crowded. Most needles were single play needles that were ground to a point which gradually wore off. On a 12-inch disc there is just as much information on a single groove near the label 5 inches in diameter and 16 inches long as there is on an outside groove 11 1/2 inches in diameter and 36 inches long. Long-wearing recording and playback needles allowed standardizing on outside-in recording. A modern needle or stylus fits the groove as shown in Figure 12.1.

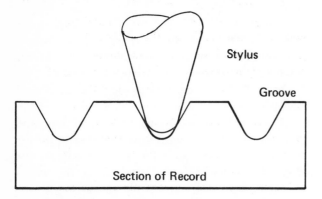

FIGURE 12.1. Stylus and Record Groove

A section of stereo modulation on a disc is shown in Figure 12.2. Both the frequency and amplitude of various sounds can be easily seen.

Disc records for many years were 10 and 12 inches in diameter providing a maximum of 5 min-

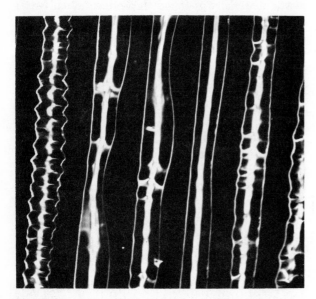

FIGURE 12.2. Photomicrograph of Record Grooves

Courtesy of Shure Bros.

utes per side. About 1930 both motion pictures and radio demanded fifteen minutes of uninterrupted sound, and the electrical transcription resulted. This record, usually called a transcription, was 16 inches in diameter and revolved at thirty-three and a third revolutions per minute. Both vertical and lateral modulation and outside-in and inside-out directions were used so that it was necessary to read the label for playback instructions. Sound for motion pictures very soon changed from discs to an optical track on the film. Radio and other non-home sound requirements were met by transcriptions for twenty-five years, and they are still used in education where libraries of them were assembled some years ago. The transcription requires a tone arm about 12 inches long as well as the slow speed.

In 1948 the long-playing (LP) record was released along with a specially designed record player. By drastically reducing the needle size, groove size (microgroove), groove spacing and adopting the transcription speed, it was possible to record up to twenty-five minutes per side. New lightweight pickups, semipermanent needles and tough vinyl plastics made it possible to obtain higher fidelity and much reduced surface noise. The LP record was supposed to obviate the need for record changers which had proved expensive and unreliable.

In 1949 the 45 record and a compact, simple and reliable changer were released in order to provide up to fifty minutes of sound without attention. The pickup, needle and groove parameters were similar to the LP system, but the speed was forty-five revolutions per minute; the diameter was 7 inches and the center hole was 1 1/2 inches.

After a few years of compatibility problems, most record players would operate at three speeds with two needles, and changers would change a stack of any one configuration. Schools often purchased record-transcription players that also had a long arm to handle 16-inch records. Schools have seldom purchased record changers.

Speech does not require nearly the frequency and volume range of music so a sixteen revolution per minute "talking book" disc was developed. It looks exactly like the 45 and uses a regular microgroove needle. Most school players now provide this slow speed, but it is seldom used.

The physical characteristics or parameters of the five discs are shown in Table 11. The first two are seldom used anymore, and the last one has not gained much popularity. An eight revolution per minute disc is being developed for very long-playing speech purposes.

Stereo records are now very common for home entertainment but as yet seldom used in schools outside of secondary music appreciation rooms. Stereo recording requires two microphones (usually about 6 feet apart), two amplifiers and a disc cutter that can move vertically for one channel or ear and laterally for the other channel or ear. In practice the cutter is operated at an angle to record one channel on one wall and the other on the opposite wall in a complex manner. The stereo playback needle must be free to move in both directions, detect the two vibrations, send the signals to two separate amplifiers and reproduce the sounds over two loudspeakers about 6 feet apart. Stereo earphones may be used instead of the loudspeakers.

The motion of the needle following the wavy grooves of the record is translated into alternating current electricity of very low power by the piezo-electric (crystal) effect or the interaction of a coil and magnetism (dynamic). Other systems are infrequently used. The least expensive and simplest system is to connect the needle to a crystal or ceramic element that will produce electricity when vibrated. This system also produces relatively high voltages and constant output over normal frequency ranges so that simple amplifiers can be used. The disadvantages may be lower fidelity (not necessarily so) and the crystals may deteriorate, particularly at high temperatures. Ceramic cartridges are similar to crystals, except for lower

electrical output and better high temperature resistance. Dynamic pickups employ various systems for translating needle motion into coil-magnet interaction to produce electricity. They are capable of very high fidelity, and they are normally rugged and immune to temperature. On the other hand, they usually require high gain and tone compensation amplifiers. They must also (unlike crystals) be protected from stray electrical or magnetic fields which might introduce hum into the system. Stereo pickups must have two pickup elements within the cartridge to detect the two separate motions from the needle.

Tone arms often have an adjustment to vary the weight of the needle in the groove. It should be just enough to keep the needle in the groove. Any extra weight only wears both needle and groove. Modern pickups track well at about 5 grams or one-sixth ounce.

The wires from the pickup (needle and motion-to electricity transducer) are normally shielded by having a metal braid over the wires to reduce noise that might get into the amplifier.

An amplifier uses the very small electrical signal from the pickup to control power from an electrical outlet or battery and apply it to a loudspeaker. It may use tubes or transistors and associated electronic components.

The amplifier must provide amplification or gain, adjusted with a volume control, and power output to drive the loudspeaker or headphones which convert electricity back into sound. It must have fidelity to make intelligible and pleasing sound. It may have additional inputs for micro-

TABLE 11. Physical Characteristics of Records

	STANDARD	TRANSCRIPTION	LONG PLAY	"45"	"16"
Speed in RPM	78	33	33	45	16
Grooves per Inch	110	110	250	250	400
Maximum Diameter	12"	16"	12"	7"	7"
Maximum Playing Time per Side	5 min.	15 min.	25 min.	8 min.	30 min.
Needle Tip Radius	.003" (standard)	.003" (standard)	.001" (Microgroove)	.001" (Microgroove)	.001" (Microgroove)
Center Hole Diameter	1/4"	1/4"	1/4"	1 1/2"	1 1/2"

phone and tuner and additional outputs for public address and recorder. It may have one or more tone controls to adjust the frequency response. If stereo, then it must have two of everything.

The turntable is rotated by a high-speed electric motor through a rubber tired intermediate speed wheel called an idler, or drive belt. The motor shaft ordinarily has several diameters, one for each speed, and a change mechanism to place the tire against the proper step on the shaft. Constant speed under any circumstance is very important for good sound. Some systems provide for variable speeds between the fixed speeds in order to make slight tempo adjustments. This is often accomplished by using a slightly fast motor-idler system and using a variable magnetic brake to slow it down as desired.

Selection

Record-transcription players are available with a great variety of characteristics for a variety of applications. The features of the machine should match educational needs. Features not needed often add to complexity and cost and may even interfere with regular use. Special purpose machines might better be borrowed for special occasions rather than used for ordinary purposes.

The size of the largest disc to be played determines the length of the playback or tone arm. Arms are made for 7-, 12- and 16-inch discs. The larger ones will play the smaller sizes but not vice-versa. The overall size of the machine also increases with the length of the tone arm. The diameter and weight of the turntable often increase with length of the arm although this is not necessary. A more powerful motor may also be desirable to rotate larger discs.

Four speed (16-33-45-78 RPM) machines are now most common in order to play any common record. High fidelity models may be limited to one or two speeds in order to make a less complex and more accurate drive system resulting in lower wow (slow changes in speed) and flutter (rapid changes in speed). The least expensive drive motors contain two poles, better ones contain four poles, and the best ones use synchronous or hysteresis motors in order to reduce wow, flutter and other undesired changes in speed. Variable speed controls may be desired for special purposes, but most often it is desired to play the record at exactly the speed used in recording.

Some means for separating the motor shaft from the rubber idler during periods of no use is essential in order to prevent the formation of an indentation that will result in an annoying thump with each revolution.

Machines are available with single or dual needle pickups. If both standard and microgroove discs are to be regularly played, then the extra cost and complexity of a two needle system is needed. As most educational recordings are now microgroove, a single needle will usually suffice. The microgroove needle will play a standard groove disc without harming either, but the high-frequency response will be poor since the small needle will wander around in the large groove. Universal needles with an intermediate size tip have been made to play both grooves, but they should be avoided since they fit neither groove properly. Microgroove needles are often painted red, universal needles painted green and standard needles not painted. It is unfortunate that the term *standard* has come to designate the least common present-day disc parameters. Since most new discs are stereo, it is important to use a stereo or compatible pickup even though monophonic reproduction results.

Metal, synthetic sapphire and diamond tipped needles are available for infrequent, regular and long-term use. Diamonds are recommended for heavy-duty use on microgroove records, particularly if regular needle inspection is not provided.

Varying degrees of portability are needed for various applications from built-in units for library carrels to operation outdoors under a tree. Size and weight are obviously important if a machine is to be carried about. Some means of holding or anchoring the tone arm is needed in order to prevent stylus damage during movement. Battery-operated models are available for use where no power is available, but motors require considerable power from batteries which may become weak or exhausted when most needed. The expense and bother of batteries must also be considered.

Transformer-powered amplifiers are desirable but more expensive and heavier. The transformer ordinarily makes a safer unit, and it is necessary for high fidelity and low background noise. It is also important if any other units such as tuners or recorders are to be plugged into the record player. There have been remarkable improvements in fidelity or the quality of reproduced sound in recent years, but high fidelity still requires weight, size and money. Sound level in a classroom should be measured in decibels, but that measurement is seldom possible. The volume or output of the record player is determined by the power output of the amplifier measured in watts and the efficiency of the speaker used. Both of these vary so much in actual values and methods of measurement that comparisons from published specifications are seldom possible. Sound output will generally increase with the continuous sine wave power in watts, speaker magnetic flux, speaker diameter and baffle size. Power needs rapidly increase from home to classroom to auditorium. The frequency response and decibel deviation over that response for a specified power output (usually one watt) are also important. Unfortunately, the published figures usually refer only to the amplifier and omit the loudspeaker which is equally important. Wow and flutter, measured in per cent of speed change, are important for pleasing sound. A single tone control may be provided to shape the frequency response curve in various ways or separate bass (low frequency) and treble (high frequency) controls may be provided to reduce, or both expand and reduce by about 15 decibels, either end of the response. Rumble and scratch or hiss filters may be available.

Mechanical devices may be included for automatically shutting off the motor at the end of the record, for stopping the record instantly at any desired point (called a pause control), for reversing or back spacing the record to hear a passage repeated and for cueing or placing the needle accurately in a predetermined spot.

Automatic record changers are not included in this discussion since they are seldom needed for educational purposes. However, most of the discussion would be applicable.

Provision for plugging in accessory microphones, tuners and the like may be desired for amplification and reproduction. Several different jacks are used, and the jacks and plugs must match or be supplied with adaptors. If both records and the plug-in device are to be used simultaneously, then a second volume control must be provided for two channel operation.

Provision for plugging in external loudspeakers is often provided in order to use a better loudspeaker than the one that can be accommodated within the case. The external loudspeaker can also be better placed in the room than in the machine.

The external loudspeaker may be ordered with the player or any speaker with similar impedance (3, 8, 16 ohm are most common) and a matching plug may be used. If incorrect matching is used, volume and tone may suffer but neither machine nor speaker will be harmed. Earphones and earphone distribution systems may also be plugged into the external speaker jack. A tape recorder can also be plugged into a special output jack on some players or into the external speaker jack. When the external speaker jack is used to feed a recorder, a resistor of the speaker impedance across the line may improve fidelity and protect the amplifier. This resistor may be essential for some transistor models and the instructions should be consulted if this is contemplated.

Some provision must be made for the large center hole of 45 and 16 RPM records. This is most often a disappearing plastic device permanently attached to the center of the turntable. Otherwise an accessory ring must be stored and applied as needed. Snap-in center hole adaptors are also available.

Some record-transcription players are so complete and capable that they qualify as public address systems considered in the previous section.

Maintenance

Record-transcription players require more maintenance than any other machine.

The needle or needles and associated mechanism need regular inspection and maintenance for good results. As the needle follows the groove in the record, it tends to sweep up any dirt or lint from clothing and accumulate it until it may interfere with proper operation. A soft brush can be used to brush the needle area regularly. Sometimes blowing will do the job. The common practice of rubbing a finger or thumb over the needle for cleaning and checking operation is to be avoided since it can easily bend or displace the microscopically adjusted mechanism.

The needle tip should be inspected regularly with a good magnifier (special models are made for the purpose) to be positive that it has not been broken, chipped or worn flat. Replacement needles are available for all makes and models of cartridges and in most cases are simply pushed into place. Instructions come with the replacement needle. The cartridge can in most cases be removed by taking out two small screws if this becomes necessary. Turnover cartridges tend to wear out the small wires attached to them or pull off the contact clips. Soldering may be necessary.

The weight (often mistakenly called pressure) on the needle should be checked with a special scale, if one is available, and set to the amount specified in the instructions with it. The adjustment may be on the top of the arm, but more often a screw and spring is located under it. If a good needle skips or repeats grooves when operating on a level surface, it probably needs slightly more weight. The commonly taped coin or paper clip on the top of the tone arm indicates that this adjustment has not been made.

The movement of the arm should be completely free and easy. Sometimes the lead wires that go down through the tone arm and pivot interfere and need rearrangement. A small amount of oil or vaseline may be needed. Any of the mechanical devices for shut-off, cueing, pause and so forth may interfere with arm movement if they are not operating or adjusted properly.

The turntable can be removed for inspection of the drive mechanism simply by lifting or by first removing a C-shaped retainer over the center spindle. If the turntable sticks to the spindle one person should hold the turntable up slightly with two hands while another taps the spindle downward with a piece of wood or plastic such as a screwdriver handle. The bearings of idlers and the motor may need a small amount of oil, but it is absolutely essential that no oil get on the friction drive spindle or rubber tired idler, or belts if they are used. If the rubber tire shows signs of wear, it may be rubbing on something or not matching the various steps on the speed change spindle of the motor.

The speed change pulley on the motor shaft can be loosened and moved slightly up or down. If the rubber tire has become glazed so that it slips, it may be roughened with fine sandpaper. Solvents are available for cleaning all friction surfaces including the inside of the turntable rim.

If the speaker is removable, then its cord and jack may need attention since such cords are often pulled, stepped on or tripped over. Wires may need to be resoldered inside the jack. A speaker, cord and jack can be tested for continuity by momentarily applying a flashlight battery to the two jack terminals and listening for a click.

Most damage to record players occurs when they are moved without first anchoring the tone arm. This anchoring mechanism should be properly adjusted for easy use, and a special instruction about its use may need to be attached to the machine.

When the vibrating loudspeaker is close to the vibration-sensitive pickup and the volume turned up, a very annoying low frequency mechanical feedback may occur. If the speaker is built in the box, then the only solution is to reduce the volume. If the speaker can be removed, or an accessory speaker can be used, then putting it on a separate stand or table will remove the difficulty. Dancing may sometimes vibrate the floor enough to interfere with a pickup, particularly if the whole machine or motor-pickup board is not spring mounted.

Hum and shock problems with inexpensive record players can often be corrected or improved simply by removing the power cord plug and rotating it 180 degrees and reinserting it. This puts the chassis and tone arm on the grounded side of the building electrical system.

Assignment X. Checked _____

Name _____

Date _____

RECORD-TRANSCRIPTION PLAYER

1. Brand and model _____ list price _____

2. Power requirements: volts _____ watts _____ battery _____

3. Is it transformer operated? _____ UL approval? _____

4. Maximum record diameter _____ speeds 16 _____ 33 _____ 45 _____ 78 _____

5. Can speed be varied from fixed speeds? _____

6. How is the motor shaft retracted from the rubber drive wheel? _____
   _____

7. Needle(s) or stylus (styli) provided? standard _____ universal _____ micro _____

8. How are the needles identified? _____

9. Can weight on the needle be adjusted? _____ how? _____

10. What mechanical features are provided? Automatic stop _____
    pause control _____ reverse _____ cueing _____

11. Provision(s) for microphone or other input(s) _____

12. Can microphone and record be used simultaneously? _____

13. Describe the tone control(s) _____

14. Number and diameter of loudspeaker(s) _____ Can loudspeaker(s) be detached for special place-
    ment? _____

15. What provisions are made for additional speakers, earphones or recorders? _____
    _____

16. Primary use for machine? individual _____ small group _____ classroom _____
    auditorium _____

chapter 13

Tape Recorders

Background

During the thirties and early forties enterprising and mechanically adept educators began to make use of disc recording and instantaneous playback for many school purposes. Portable machines were developed and plastic coated discs of paper, metal or glass were used. A very sharp cutting needle or stylus was vibrated and gradually moved over the surface as the disc was rotated under it. The result was a modulated spiral groove explained in the previous section. Many difficulties limited the wide application of sound recording by the disc cutting method.

Magnetic recording is not a recent invention. Demonstrations of the technique were conducted before 1900. As electrical amplification became available, there was considerable development of magnetic wire recorders for limited speech response, such as dictation. During World War II wire recording was further developed, but it was impossible to equal the sound quality of the better disc recorders.

The relative convenience of wire recording led to a brief period of popularity in education just after World War II. A 3-inch diameter spool containing more than a mile of stainless steel wire about the thickness of human hair was drawn past a grooved head with a magnetic field corresponding to the sound to be recorded. A permanent magnetic pattern was produced on the wire. The wire could then be rewound and again drawn past the grooved head which was this time connected to the amplifier input. The movement of the varying magnetic fields from the wire induced an alternating current in the coils of the head which was then amplified and fed to a loudspeaker. Since no cutting or removal of material was involved, much less attention and technical skill was required to make a sound recording. Recording time was increased from a few minutes on a disc to an hour on wire without interruption. The wire, unlike the disc, could be erased automatically by the recorder just before a new recording was made so that nonpermanent recording became very inexpensive. There was little danger in harming the magnetic pattern on the wire compared to harming the soft plastic coating of the disc.

Wire recording also had disadvantages. Even at a speed of 2 feet per second it was impossible to attain the fidelity of disc recording. Background noise, high-frequency response and echoes caused by storing one layer of wire directly against another, were particular problems. Rewinding a mile of very fine wire at high speed sometimes resulted in what was called a bird's nest. Broken wire could be tied together, but with considerable difficulty.

At the end of World War II magnetic tape recorders were found in Germany that produced much better results than wire recorders. Tape recording was developed rapidly in this country and soon became the dominant method. Simple, low cost and high fidelity machines have made the instantaneous recording and playback of sound for education very commonplace.

Schools use one-fourth inch recording tape almost exclusively, although professional recording may be done on other widths. The tape was once made of paper, but plastic about one-thousandth an inch thick is now used. A very thin layer of very fine iron oxide is bound to one side of the tape, giving it a dull appearance. This oxide coating can be magnetized, stored and demagnetized at will.

The typical school tape recorder contains a microphone to convert sound into alternating current electricity. This weak electrical facsimile of the sound is amplified and applied to a recording head shown in Figure 13.1. The recording head has a coil of wire around an iron core to convert electricity to magnetism. The iron core has a very narrow gap across which the magnetism is concentrated. The tape with randomly oriented magnetic particles is drawn over and in close proximity to the gap. During this recording process the magnetic particles are oriented according to the frequency and intensity of the magnetic field across the gap. The recorded tape is wound on a take-up spool and then rewound prior to playback.

During playback the tape is again drawn over the recording head, but this time the coil of wire around the iron core is connected to the amplifier input instead of the microphone. The magnetic patterns moving over the gap induce a small electric current in the coil which is amplified and connected to a loudspeaker to reproduce the original sound. Special machines may have separate record and reproduce heads but the combination head is most common.

In order to reuse a tape it is necessary to erase the previous recording. All school tape recorders erase automatically whenever the machine is put in the record mode. Two buttons or devices must be operated in order to prevent accidental erasure. Erasure of the tape is accomplished by drawing it past a special erase head just prior to the recording head. The erase head contains a coil, core and

wide gap supplied with a high-frequency alternating current which saturates the tape with magnetism and then leaves it randomly arranged for a new recording to be made.

The tape movement during recording and playback must be very constant in order to maintain pitch and timing and to avoid unpleasant wow and flutter. This is accomplished by driving a heavy capstan and flywheel with a constant speed motor and holding the tape firmly against it with a rubber pinch or pressure roller. The tape is usually pressed against the record-playback head with felt pressure pads to assure good contact.

Several tape speeds are in common use. When tape recording first became popular, it was necessary to move the tape at 15 inches per second in order to obtain true high fidelity reproduction. With narrower gaps and finer oxide particles, it is now possible to get good response to the high frequency limits of human hearing at 7 1/2 inches per second so that this is the highest speed usually found on school tape recorders. Speech ordinarily does not require such high fidelity, and a speed of 3 3/4 inches per second is most used. The slower speed permits twice the recording time on a given length of tape. Most school tape recorders have these two speeds available with a simple speed changing device. It is of course necessary to play back any recording at the same speed at which

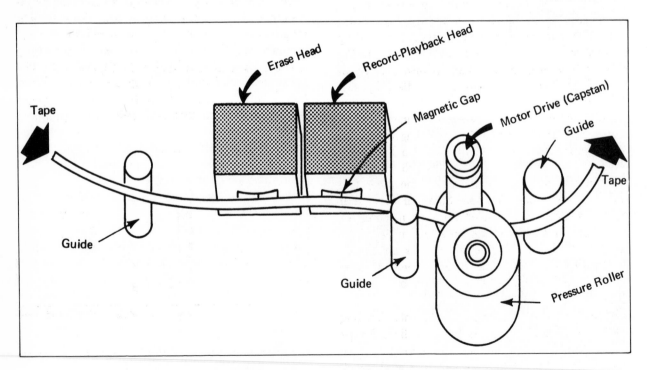

FIGURE 13.1. Tape Recorder System

it was recorded. Slower tape speeds are becoming available in order to save on tape costs and in order to make very small and portable machines. Speeds of 1 7/8 and 15/16 inches per second are used for these purposes. Although they are not high fidelity they may give surprisingly good sound.

Most school tape recorders are described as "reel to reel," which means that tape must be pulled from a supply reel, threaded through the machine and attached to an empty take-up reel before the record or playback mechanism is operated. Threading is usually accomplished simply by dropping the tape in a long slot and starting it around the take-up reel. Cassette or cartridge machines are becoming popular for portable, automobile, home and some school uses. With these the tape is permanently enclosed in a plastic shell which is pushed into or pulled out of the special machine. There are several noncompatible cartridge machines.

The original tape recorders and most modern professional tape recorders have one wide track down the center of the tape. These are called full track or one track machines. They provide highest fidelity and no confusion. Most school recorders are two track machines in which one recording can be made in one direction along one half of the tape width. The tape, without rewinding, can then be taken off the take-up spindle, turned over, put on the supply spindle and run through the mechanism a second time to make a second recording or a continuation of the first. Dual track recording reduces tape costs by one half. It also may result in confusion. One track cannot be edited by cutting and splicing without ruining the other track. Some commercial tapes have the identical recording on both tracks so that no rewinding is needed. Stereo tape recorders most often use a four track format in which two one-fourth width tracks are used in one direction followed by the other two tracks in the other direction. Some of these four track machines can be used for four monophonic recordings or special sound on sound multiple recordings. Various track configurations are shown in Figure 13.2. The actual tracks cannot be seen unless coated with a fine suspension of iron particles.

There is often confusion about interchanging various track tapes and recorders. A full track tape can be played on any mono or stereo machine with the proper speed. If it is turned over at the end,

the recording will be played backward. A multiple track recording played on a full track machine will have all tracks, both forward and backward, played simultaneously. If it is desired to have a school tape used at a radio station or at a commercial recording studio, a new tape or a completely erased tape (bulk erasers are available) should be used in one direction only at 7 1/2 inches per second.

Full Track

Half Track

Quarter Track

FIGURE 13.2. Tape Recording Configurations

A complex electronic system is necessary to produce good recording and playback. The magnetism resulting in a medium such as iron oxide from a magnetizing force like that produced by the recording head is not a linear relationship but rather a curve such as that shown in Figure 13.3. The magnetization of the tape is slow to start, A to B, then proceeds in almost a straight line, B to C, and then tapers off so that additional force produces

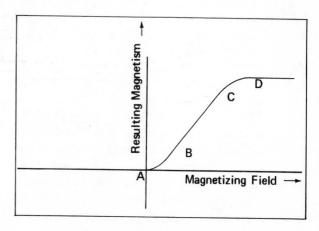

FIGURE 13.3. Magnetization of Tape

no more magnetism, C to D. High fidelity can result only from magnetism along the straight part of the curve from B to C.

To keep all recording above the lower curve, a steady and supersonic signal (about 50,000 Hz) called bias is applied to the recording head. In practice this is derived from the erase signal. The use of bias also improves the signal to noise ratio.

To keep from reaching the upper curve and saturating the tape, some form of volume level indicator must be used and observed carefully during recording. A meter called a VU meter is used in all professional work and a small version is used on some school units. The needle should peak near zero (about two-thirds the way up) on the professional meter or stay in the green area of many small meters. Live sounds vary so much in intensity that setting the level must be a compromise between distortion due to overmodulation and losing the signal in the background noise during low level periods. Professional studios and radio stations use complex automatic volume leveling devices, and some newer tape recorders have simple versions to help adjust the volume automatically. Magic eye tubes with a variable shadow in a green window are often used. It will be necessary to consult the instructions with the machine to determine how the shadow should look on the highest permissable levels. If no instructions are available, the volume should be adjusted to keep in the middle of the shadow range. Neon flash lamps, often in pairs, are also used to indicate proper volume control setting during recording. If two lamps are used the one marked "normal" should flash regularly and the one marked "distort" should seldom if ever flash.

The necessary mechanical-electronic functions may be controlled with push buttons, rotary knobs, a "joy stick" or some combination. A simplified tape recorder diagram is shown in Figure 13.4.

The record-playback-fast forward-stop-rewind switch or switches are made up of several sections that operate simultaneously after the tape has been threaded. In the record position (plus a safety release button often colored red) the microphone is connected to the amplifier input, the loudspeaker is disconnected (unless a special switch labeled "monitor" is operated), the erase and bias oscillator is actuated, the tone control is disconnected, the amplifier output is connected across the record (actually record/playback) head and the volume level indicator, the pinch or pressure roller presses

the tape firmly against the motor driven capstan (the motor usually runs all the time the machine is on), the pressure pads press the tape against the erase and record heads, the supply reel has gentle braking applied to prevent spillage and the take-up reel is driven through a friction clutch to roll up the tape. In the stop position, brakes are applied to both reels, and the pressure pads and pinch roller are released. Fast forward and rewind positions release both pressure pads and pinch roller, release the brakes and transport the tape rapidly in either direction. Some machines also lift the tape away from the heads during fast forward or rewind to eliminate the peculiar sound called "monkey chatter" and to reduce wear on the heads. In the playback position the microphone is disconnected, the playback head is connected to the amplifier input, the erase and bias oscillator is deactivated, the amplifier ouput is connected to the loudspeaker, the tone control is activated, the pinch roller and pressure pads press against the tape, the supply reel is gently braked and the take-up reel is driven. A pause control on many machines can be used to stop the tape momentarily.

A tape position indicator consisting of three digits is usually provided. The digits can be quickly reset to zero with an adjacent thumb wheel. The figures are arbitrary, rather than feet, inches or seconds. It is possible to make a record of where various sounds are on a tape and proceed quickly to the desired place by using fast forward and rewind. The numbers do not mean the same thing on different machines or models.

Although most school tape recorders have only one input channel, there are usually two input jacks for different signal levels. One jack is for a high impedance record player, radio tuner, or the output from another tape recorder. The other jack is for the microphone which needs considerably more amplification than the other signal sources. If a low impedance microphone is used, a special line transformer is needed. Professional recorders accept low impedance microphones directly. Much better results are obtained if sound from a record player, radio or another tape recorder is patched directly into the input with a special cord called a patch cord rather than using the loudspeaker and microphone. Most machines now have an output jack labeled external speaker, amplifier, earphones, recorder and so forth to provide a signal to the patch cord. It is necessary to have proper jacks or jacks plus adaptors on each end. The wire is

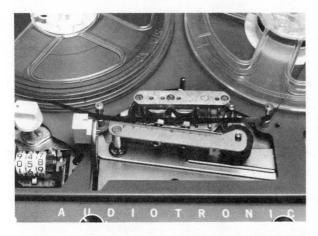

A. Open for threading.   B. Ready for recording, but with covers removed.

Courtesy of Audiotronics Corp.

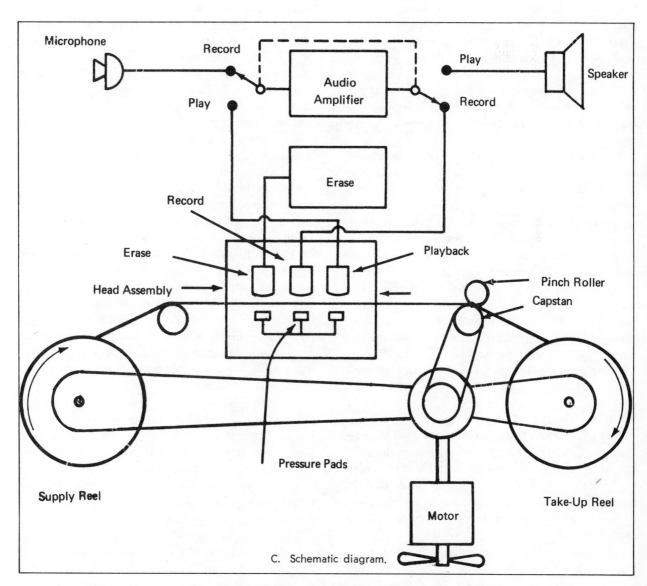

C. Schematic diagram.

FIGURE 13.4. Tape Recorder Mechanism

shielded to reduce hum and other noise. If there is sparking, a shock or hum from connecting older machines that do not have self-grounding AC plugs, the cords should be plugged into the outlet in different combinations. If the signal source does not have an output jack, a patch cord with alligator clips on one end can be connected to the two soldered voice coil connections on the loudspeaker. It may be necessary to reverse the connections for best results.

If recording is done from other than a nearby microphone, it is often desirable to monitor or listen to the material while it is being recorded. This can often be done by operating a switch marked "monitor" or "public address." This also permits the recorder to be used as a public address system with the usual feedback problem if microphone and loudspeaker are near each other and the volume control advanced. To avoid feedback or squealing, the monitor is usually left in the off position.

Most school recorders will not permit mixing two or more sound sources since they have but one channel. Accessory mixers can be plugged into the input jack.

The loudspeaker included in the case usually has very limited capabilities compared to the rest of the machine. An accessory loudspeaker in a baffle can usually be plugged into the external speaker jack. A better microphone than the standard one provided will also improve recordings.

It is sometimes desired to erase a tape without making a new recording. This is easily done by running the tape through the machine in the record position with the volume control set at zero. By carefully watching the counter it is possible to erase or edit a selected portion of a recording. The back of a tape can be marked for identification with a plastic marking pen such as those used with overhead transparencies.

Selection

The popularity of tape recorders is indicated by the large number of brands and models on the market. Most school machines are portable reel to reel, two or three speed, two track monophonic models. Battery portable, cartridge and stereo machines are much less used in the classroom and not considered here in detail.

Compatibility should be considered if there is a possibility of using commercial tapes or exchanging tapes with another person. Most school tape

reel to reel half-track recorders will accommodate 7-inch diameter reels at 3 3/4 and 7 1/2 inches per second so that almost any tape will play on almost any machine. If an odd machine is purchased it may be necessary to rerecord material before use on a common school machine.

Weight and size should be considered if portability is needed. Most machines weigh about 20 pounds due to the heavy motor and power transformer. Lighter machines may not produce the other qualities desired. As more machines are left in the places where they are regularly used, portability becomes less of a problem. A battery portable can be used in combination with the classroom model with the help of a patch cord.

The maximum reel size should normally be 7 inches. Three and five inch reels are used for some home and special purpose machines. A 7-inch reel will hold 1200 feet of standard thickness (.0015" or 1 1/2 mil.) tape and record in one direction for one hour at 3 3/4 inches per second. The capacities of various reel sizes for various thickness tapes and the resulting recording times at various speeds and numbers of tracks are indicated in Table 12.

Tape selection is also a problem. Standard thickness tape is usually made of an inexpensive plastic called acetate. It is easy to handle and does very well for all normal purposes. Thinner tape enables more length to be put on one reel, but a stronger plastic such as mylar or polyester must be used, at increased cost. The thinner tape may be more difficult to handle. Tape for continuous use, such as in a language laboratory, may have a very thin coating over the oxide to reduce wear. Tape comes in various colors and on various colored reels if identification is a problem. High output, low print-through and low noise tapes are needed only for special applications. Practically all tapes will play on practically all machines found in the school market.

The usual speeds are 7 1/2 and 3 3/4 inches per second so one of these speeds should be selected if any tape exchange is expected. If very long recordings are needed with a sacrifice in sound quality, then a machine with a slower speed can be purchased. The same result may be obtained with no loss in quality but at a higher tape cost by using thinner tape.

The frequency response reported in the literature tells the lowest and highest frequencies that can be recorded and reproduced by the machine with the number of dB variation, such as 40-10,000 Hz ± 3

TABLE 12.   Recording Time for Tapes

DIAMETER (Inches)	THICKNESS (Mils)	LENGTH (Feet)	SINGLE TRACK SPEED (ips)			DUAL TRACK SPEED (ips)		
			1 7/8	3 3/4	7 1/2	1 7/8	3 3/4	7 1/2
3	1.5	150	15	8	4	30	15	8
	1.0	225	23	12	6	45	23	12
	0.5	300	30	15	8	60	30	15
5	1.5	600	60	30	15	120	60	30
	1.0	900	90	45	23	180	90	45
	0.5	1200	120	60	30	240	120	60
7	1.5	1200	120	60	30	240	120	60
	1.0	1800	180	90	45	360	180	90
	0.5	2400	240	120	60	480	240	120

(Body of Table Indicates Minutes)

dB, for a given speed. Unfortunately, there is no standardization in the way these figures are measured and reported. The microphone and loudspeaker ordinarily are not included in the measurement. Some figures only report the playback amplifier characteristics. In comparing machines, it is essential to compare figures measured and reported in the same way, and this may be impossible to do at the present time. Reported frequency response is not much of an indication of tape recorder overall sound quality.

Wow and flutter indicate that the tape is moving through the mechanism at other than constant forward speed. They are kept low by heavy motors and flywheels, good mechanical design and so on. The combined wow (low frequency changes) and flutter (high frequency changes) are reported as a percentage, generally less than 0.5 per cent for school machines. The power output of the amplifier is given in watts, but unless the method of measurement, frequency response and distortion are included with it, practically no meaning can be obtained.

Most tape recorders have more than adequate power to drive their built-in speakers. If one is to be used with an external speaker for a large group then a machine with 10 watts EIA (Elec-

tronic Industries Association) or continuous sine wave power at no more than 5 per cent harmonic distortion would probably be adequate.

Signal to noise ratio indicates how much louder the sound it is supposed to produce is than the noise it is not supposed to produce. Good machines have figures for this around 45 dB and higher. Unfortunately, they do not usually include the mechanical noise of the mechanism which may interfere with listening close to the machine. If the machine is very quiet, as it should be, it may need a good pilot light to remind someone to turn it off at the end of the period or day.

A pause control may be desirable in order to stall the tape (but not the motor or flywheel) while a comment is made. It should work easily and stop the tape without making odd sounds that might distract the listeners. Pause controls often can be connected to accessory foot controls so that a speech can be transcribed by a secretary. Pause controls are also used to eliminate unwanted sounds during recording.

Enough inputs and outputs should be available to feed in and out any signals desired. One or more patch cords may be included or available as accessories. Higher quality microphones and loudspeakers will improve the sound. They may be purchased to match the machine or a technical person can easily make good selections. Earphones or sets of earphones are often used with tape recorders. Most any earphones with the appropriate plug to match the recorder jack may be used.

Automatic head demagnetization is included on some machines to remove any residual magnetism that may accumulate in the record/playback head. If this feature is not included, an external demagnetizer may be used if necessary during maintenance.

Automatic stopping or shut-off at the end of a tape, or if the tape breaks, is included on many machines. It may mean an extra step in threading. It is very desirable if a machine is to be left in operation and unattended for any period of time.

Power consumption of most machines is around 100 watts and of no concern unless many machines are to be used on one circuit, such as in a language laboratory.

Maintenance

The erase and record/playback heads normally accumulate a good deal of dust and dirt from the environment and particularly from tape that has been allowed to spill on the floor during threading, editing or rewinding. As the tape goes through the machine it is pressed firmly against the heads by the felt pressure pads so that any foreign matter is rubbed off and deposited in the head area. Some of the oxide coating of the tape is also worn off and deposited on each pass through the machine. Oxide rub-off may be more severe with some offbrand or white box tapes. The tape recorder should be covered when not in use.

The head and pressure pad area of tape recorders must be cleaned very regularly in order to avoid poor erasure of previous recordings, loss of volume, loss of higher frequencies and intermittent sound. To make this easy a cover for this area will snap off, or it is removable with a small screwdriver. Some dirt can be blown out. The tape path past the heads is best cleaned with cotton twisted on a toothpick and then dipped in alcohol. Commercially prepared cotton wads on sticks for medical purposes are often used. Tape recorder maintenance kits contain large versions of these. Alcohol or special head cleaner is the only fluid recommended by most recorder and tape manufacturers. Carbon tetrachloride is *not* recommended. No metal tools should be used in cleaning the heads due to the danger of scratching the highly polished surfaces over which the tape must slide. A scratched head can wear tape very rapidly and soon interfere with the whole process. No magnetized object should be brought near the heads or residual magnetism may remain. A screwdriver used in this area should be checked for magnetism by trying to pick up a paper clip or thumbtack.

The felt pressure pads are cemented to the metal arms that move them in and out of position. They must be cleaned without disturbing them. An old toothbrush can often be used with great care. Alcohol will not loosen the cement, but many other solvents will. If one of the pads is dislodged, it is probably to be found in the immediate area and can be recemented with a small drop of plastic or epoxy cement, being very careful not to have any hardened cement exposed to the tape.

The entire tape path from supply reel to take-up reel should be inspected for cleanliness and anything that might interfere with tape travel and good contact with the heads. The capstan connected to the motor which drives the tape should be absolutely clean. This also applies to the rubber pressure or pinch roller which holds the tape against the capstan. A small piece of cloth moistened with alcohol is helpful in cleaning the entire route. If ordinary pressure sensitive cellophane tape has been used for tape repairs or left at the beginning or end of a tape, some of it or its adhesive may dislodge in the tape path. Only wood or plastic tools should be used for removing stubborn pieces.

A common tape problem may give some of the symptoms of dirty heads. Tape must be used with the oxide coating against the gaps of the heads in order to have good erasure, recording or playback. Sometimes a tape is inadvertently turned over during rewinding, splicing or threading. This is particularly apt to happen with very thin tape, or if coated tape is used so that the dull oxide coating is difficult to recognize. A broken and mistakenly spliced tape will work one side of the splice and not the other.

The recording/playback head may accumulate magnetism which will result in increased background noise on reproduced tapes. Special head demagnetizers are available from audiovisual suppliers. They contain an alternating magnetic field (from 60 Hz electricity) that is brought near the tape path past the head and slowly moved away before it is turned off. Any metal parts of the demagnitizers should not actually touch this polished tape path. Some machines have automatic head demagnetization.

The magnetic gap in the record/playback head must be at exact right angles to the tape. Adjustment screws to align the head are normally covered with colored cement to identify them and prevent accidental movement. They should be touched only by a trained person with a special alignment tape.

A top plate or cover under the supply and take-up reels can be easily removed with a screwdriver on most machines to expose the complex tape drive system. There are many bearings that must have very low friction for proper operation and many clutches, belts, idlers and brakes that must have varying amounts of controlled friction for proper operation. Attention to this area should be given only if mechanical transport problems are present and the servicing instructions are available to a competent technical person. Any oiling should be done with an oiler that has a hypodermic type dispenser to deliver a very small amount of oil to the exact point desired. Alcohol or special solvents may be used to clean friction surfaces.

The microphone and/or cord used with the tape recorder may be damaged from rough handling. Sharp noises as the microphone is moved are usually due to a bad microphone cord connection at one end or the other. This can be checked by using the machine as a public address system by means of the monitor switch. Other microphone-cord problems are best checked by trying another similar microphone.

Some recorders contain fuses which may burn out or mechanically fail, since they are deliberate weak links. If a fuse is used an exact spare should be taped nearby for possible use. If the replacement fuse burns out, the machine needs electronic servicing. Some machines use a circuit breaker which can be reset by pushing an exposed and labeled button. A defective power cord or plug can also result in an inoperative or intermittent machine.

Assignment XI. Checked _____

Name _____

Date _____

TAPE RECORDER

1. Brand and model _____ list price _____ UL approval _____

2. Power required: watts _____ volts _____

3. Maximum reel diameter _____ maximum length of standard tape _____

4. Number of tracks _____ mono or stereo? _____

5. What tape speeds available? _____

6. Weight _____ dimensions _____ pilot light _____

7. What inputs are provided? _____

8. What outputs are provided? _____

9. What two buttons or devices must be operated after the off-on switch to start recording? _____
_____

10. Is a pause, cough or edit device included? _____ how used? _____

11. Is an automatic shut-off provided in case the tape breaks? _____

12. Is it possible to monitor through the loudspeaker while a recording is being made? _____
how? _____

13. What type of recording level meter is used? _____ How is correct level indicated? _____
_____

14. Is fast forward speed available as well as rewind? _____

15. What amplifier specifications are given? Watts output _____
Frequency response _____ signal/noise ratio _____
Wow & flutter _____ Is any indication of measuring standards given? _____
_____

16. Make a rough but labeled diagram to show how the machine is threaded.

chapter 14

Motion Picture Projectors

The fundamental principle of the motion picture was discovered thousands of years ago. When an object upon which the eye is focused is suddenly removed or the light is extinguished, the image of that object remains or lingers on the retina of the eye or in the brain for a fraction of a second. The phenomenon is called persistence of vision.

If a series of perfectly still objects or images with a slight and progressive change in detail or position is presented to the eye in rapid succession, each image is stored and fused into the succeeding image. The result appears to be movement. Many illuminated signs provide apparent motion simply by progressive lighting of stationary lamps.

Early experiments with simulated motion pictures were based on observing a rapid sequence of progressively different still drawings or pictures on the inside of a drum with slots for viewing, or through a peephole as cards attached to a drum were flipped into position by a crank. Rather elaborate presentations involving thousands of cards were developed for individuals to observe through coin-operated machines. Many of these machines can still be operated in museums.

George Eastman paved the way for photographing and displaying a rapid sequence of actual pictures rather than drawings by developing flexible photographic film late in the nineteenth century. Thomas Edison made use of Eastman film to construct individual viewing machines called kinetoscopes which presented a rapid succession of still film images.

About the turn of the century several inventors, including Jenkins, Armat and the Lumiere brothers, perfected projectors so that a group could see a motion picture on a screen in a darkened room. Improvements were rapidly made and motion picture viewing in theaters became very popular.

The projection system has remained essentially the same for more than sixty years. It is shown in Figure 14.1. A long strip of film including a series of progressively different transparent positive images and a series of sprocket holes or perforations for advancing and positioning purposes is prepared with a movie camera and processing laboratory. The film is mounted on a reel and pulled off at constant speed by a sprocket or toothed wheel that engages the perforations in the film. Some sort of a lock or shoe keeps the film securely on the sprocket wheel. A projection system with lamp, condensers, film aperture and projection lens (very similar to those in still projectors) is used to recreate the brightly lighted film image on the screen. The projection elements may be small to accommodate small film images. The film must be manipulated through the projection aperture in a peculiar and complex manner. Each individual film image must be held perfectly still and in exact location while its image is projected on the screen for a small fraction of a second.

A claw operated by a complex set of cams enters or engages the film perforations at precise intervals to advance the film and locate it properly in the aperture. Between the constantly rotating sprocket wheel and the intermittent aperture mechanism, it is essential to have an upper loop of slack film (room for two fingers) in order to avoid tearing the sprocket holes. Under the intermittent mechanism there must be a lower loop of slack film about the same size and another sprocket wheel to move the film and send it on toward a friction driven take-up reel. The film is then rewound and ready for storage or reuse.

Many sizes of film and sprocket hole arrangements have been developed and promoted. Standardization is not a new problem. Films with overall

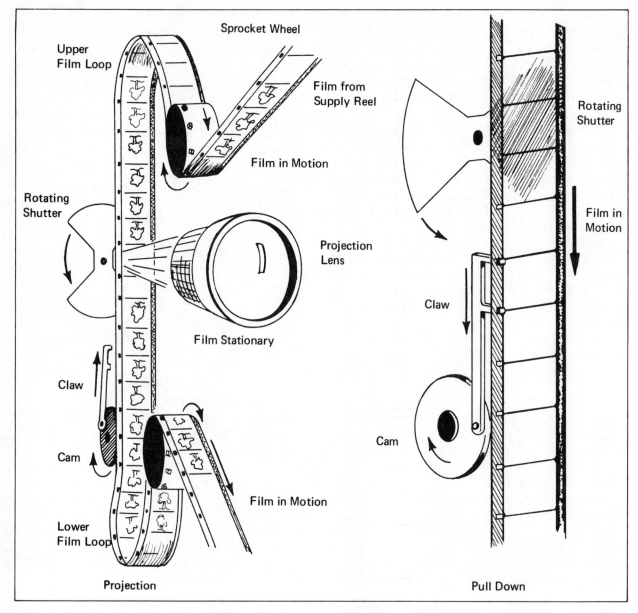

FIGURE 14.1. Motion Picture Projection System

widths of 35mm, 28mm, 17.5mm and 9.5mm were early favorites. Sprocket holes have had different shapes, sizes and locations. Even double track films similar to dual track tapes have been used.

About the time of World War I, the 35mm film now used almost exclusively in commercial theaters became the standard format in this country. It was made of a very flammable and dangerous plastic called cellulose nitrate or just nitrate. It could be shown only on approved machines in fireproof booths with many safety precautions and trained projectionists. Slow burning film stocks were developed for portable use, but for many years ni-

trate was stronger, more transparent and easier to splice. Nitrate 35mm film must be labeled regularly along the edge, kept in metal containers and never mailed. It is seldom used today, but any 35mm movie film could be nitrate and very dangerous. All filmstrips are on 35mm safety film. Schools made very limited use of 35mm film in auditoriums. A few portable machines and a few educational safety films were made, but they never became popular.

Sixteen millimeter film was developed about 1924, and it soon became a popular and standardized educational format. Most educational films are now 16mm. All 16mm film is slow burning or safety film. During the twenties several libraries of short silent educational films were produced for classroom use. Silent 16mm film had two rows of sprocket holes and double-toothed sprocket wheels and claws. A standard rate of sixteen frames or 24 feet per second was adopted.

A combination of transcription discs and 35mm film was developed for theater sound motion pictures in 1929. A very few sound on disc 16mm films were also produced. It was soon found to be too difficult to keep the disc and film in synchronism, particularly if pieces of the film had been removed during repair and splicing.

In the early thirties optical sound tracks were developed for both 35 and 16mm films. If properly produced, permanent synchronism of picture and sound resulted. On 35mm theater film, the optical sound track was accommodated beside the pictures and inside the sprocket hole area. In order to find room for sound on 16mm film, one row of sprocket holes was eliminated and a one-sided sprocket and claw system developed. Many sound films during the thirties had damaged sound tracks from projection on silent machines. Some silent 16mm projectors were developed with teeth on only one side to prevent sound track damage. Silent 16mm machines are seldom used today except for time and motion analysis. Most sound projectors will also play silent film at the new standard of eighteen frames per second. Silent and sound film are shown in Figure 14.2.

Mechanical and magnetic sound tracks have already been considered with phonographs and tape recorders. The most common film sound track is a continuous ribbon of a light and shade representation of the sound wave. It can easily be seen with the naked eye. It is produced by feeding the amplified sound signal to a light valve which varies the area or intensity of a light focused on the light sensitive film. Both variable density and variable area sound tracks are regularly used on projectors without any adjustments. A master film with pictures and sound can be copied or reproduced at high speed in special film printers.

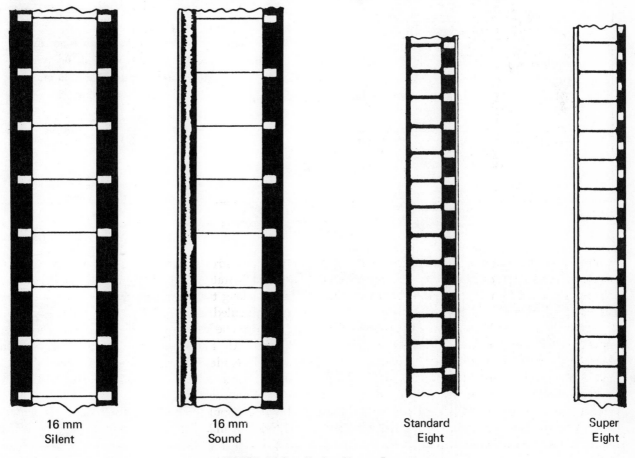

16 mm
Silent

16 mm
Sound

Standard
Eight

Super
Eight

FIGURE 14.2. Motion Picture Formats

Some films are made by exposing and developing a negative (sky dark and grass light) and then using it to print a positive copy. Other films are made by using direct positive or reversal film which is processed into projection film without using a negative. Both are capable of producing excellent results.

The light and shade of the sound track is converted into varying or modulated light by shining a bright but very narrow beam of light through it. The light is produced by a small incandescent exciter lamp and a complex but small lens system. The exciter lamp is usually operated on direct current or high frequency alternating current (about 50,000 Hz from an oscillator in the sound electronics) in order to avoid audible 60 Hz hum in the sound output. When sound on film was standardized (early thirties), it was decided to increase the rate of frame projection from sixteen to twenty-four per second in order to reduce flicker and improve high-frequency audio response. Even at twenty-four frames or 36 feet per second, a 5000 Hz sound wave requires .0014 inches of sound track for each cycle, and this is about the best that is ordinarily possible. One brand of sound projector provides a lever to focus the light slit on the sound track for best high-frequency response. Sound on film does not produce high fidelity on 16mm equipment. Signal to noise ratios of only about thirty dB can usually be attained. Thirty-five millimeter

theater sound films travel at 90 feet per minute and high fidelity sound can be produced.

The varying or modulated light from the exciter and sound track is fed directly into a photoelectric or light sensitive cell or routed to one through a mirror or lucite tube. The cell changes the varying light into varying or alternating electricity which is then fed to a high gain amplifier and then to a loudspeaker. The photocell may look like a vacuum tube, but it has no filament and does not glow. Small solid state light sensitive devices are used on many newer machines.

At the time the sound is taken from the film, it is essential that very constant forward speed be maintained to avoid wow and flutter. To accomplish this it is necessary to place the sound 7 1/2 inches ahead of the picture so that the intermittent motion can be effectively separated from constant motion. It is also necessary to press the film firmly against the hub of a heavy flywheel at the instant the sound track is scanned. A common practice is to have the sound track protrude over the edge of the flywheel hub and scan it at that point. Projector sound systems are diagramed in Figure 14.3.

If the sound and picture take-offs are not exactly 7 1/2 inches apart, the sound will not be synchronized. A threading line or automatic loop setter will help assure good sound synchronization.

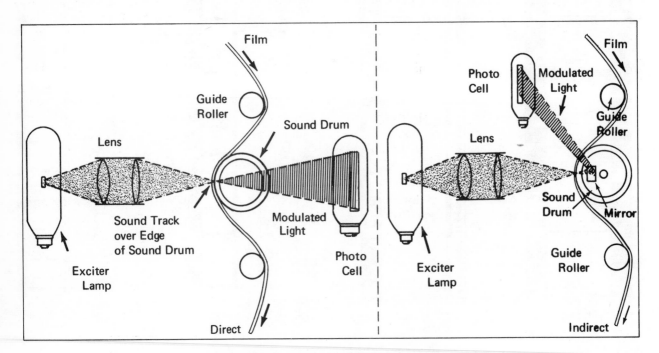

FIGURE 14.3. Motion Picture Sound Systems

After sound is obtained the film is moved along by a sprocket wheel with its lock or shoe and then guided to the friction driven take-up reel. The take-up reel must be friction driven in order to rotate rapidly when the film is going around the small hub and slowly when it is going around the full reel. In order to minimize sudden tugs on the film, a spring-loaded snubber or shock-absorbing idler is used between the last sprocket wheel and the take-up reel.

A reel of 16mm film was standardized at 400 feet, about 7 inches in diameter, and that was the largest size that most silent projectors could handle. Sound projectors were made to hold 1600 or even 2000 feet of film which may be referred to as four or five reels, even though it is on one reel. The longest sound films normally used are on 1600-foot reels which will run continuously for about forty-five minutes.

Threading diagrams for the most common manual threading sound projectors are shown in Figure 14.4. Many of these have remained almost unchanged for twenty years. It will be noted that they may vary widely, and skill in threading one projector may be of little help in threading another unless the steps that must be included in order are understood. Modern projectors are provided with much better labeled film paths and instruction booklets than older models.

Several sound projectors are now available with automatic or semiautomatic threading through the intermittent and sound take-off areas. Specific instructions for each machine should be followed.

It is difficult to obtain very bright screen images from 16mm films because of the small image on the film. An individual frame has an image .292 inches by .402 inches or an area of .117 square inches. The projector aperture is slightly smaller. To fill a 52-inch by 70-inch screen, requires an area multiplication of 32,000 times and the illumination is decreased by the same ratio.

Even at twenty-four frames per second flicker on the screen became annoying as light intensity increased through the years. To reduce flicker the light is ordinarily interrupted several times for each frame and once while the pull-down is effected. Various systems for exposing the image more than 50 per cent of the time are also used to increase the number of lumens on the screen.

Still projection of a single motion picture frame is commonly wanted by teachers, but it is a very difficult technical thing to do. The forward motion must be brought to a stop, which results in ludicrous sound. The shutter must be in the open position in the aperture. Several starts and stops may be necessary to attain this condition or a hand knob may permit manual adjustment. The heat from the lamp must be drastically reduced with an automatic filter in order to minimize film damage. Even so, the film may be overheated, and the light level on the screen will be disappointing. The distortion of the heated film often moves it out of focus to further degrade the screen image.

Reverse projection is available on some sound movie projectors in order to repeat a desired portion. This is particularly helpful in setting up and testing a machine before the audience arrives, and then backing up to the start. If used, the machine should be allowed to come to a complete stop before changing its direction. A reversible motor system is usually more complex and expensive.

Most sound projectors have a speed control to permit projecting silent films at the now standard eighteen frames per second. The film sound may be automatically or manually turned off. A microphone can be plugged into most sound projectors for adding a commentary.

Colored motion picture film became available in the mid-thirties with extra steps and extra cost. Colored film still costs approximately twice as much as black and white. It is often more difficult to splice if it breaks. There is little evidence that colored film results in more education unless color is essential to identification, but there is a widespread preference for color.

Eight millimeter film also became available during the mid-thirties in order to reduce the cost of movies for home or amateur use. These films were designed for small groups of people to view at night. This size film is now called regular or standard eight.

In the early fifties, magnetic sound on 16mm film was developed in order to obtain higher fidelity sound and to permit adding or erasing and rerecording sound in the field. Sound movie projectors with both optical and magnetic sound systems were and are made. They have enjoyed only limited popularity.

In 1956 the recording and reproduction of both sound and pictures on 2-inch wide magnetic tape was perfected. Television tape recording is discussed in the section on television. Classroom motion pictures may one day be reproduced on tele-

FIGURE 14.4. Threading Diagrams for Common Manual Sound Projectors

RCA

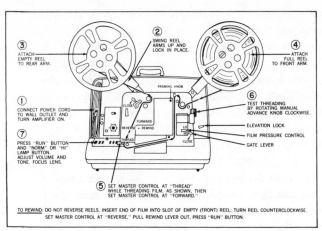

Graflex

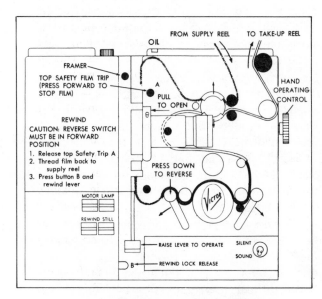

Kalart/Victor

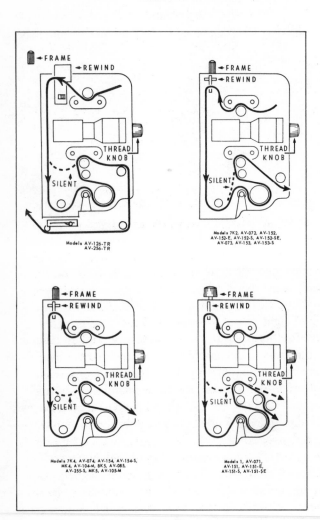

Kodak

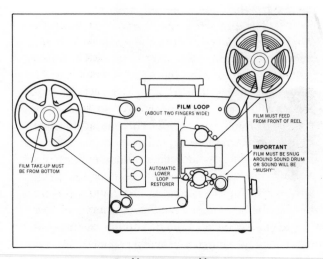

Bell & Howell

vision receiver-type machines from tape cartridges in lighted rooms.

In the middle sixties several independent 8-millimeter developments were aimed at putting regular 8-millimeter film in cartridges for automatic threading and continuous loop operation, adding magnetic sound to regular 8, reducing sprocket hole size to permit slightly larger pictures and much wider sound track and reducing sprocket hole size to permit much larger pictures and a usable sound track. So many 8-millimeter formats are in use that considerable confusion results. Until some standardization occurs it is necessary to match the film (sound and picture format) and its container (reel or cartridge) to the machine. There seems to be a widespread and definite move toward the "super-8" format at this time. Regular 8 (R8) and super-8 (S8) films are compared in Figure 14.2. Since many millions of R8 films have been made, machines for projecting them will be around for a long time. Several rather complex reel to reel projectors for using either format have been developed for users who have need to project both regularly.

The super-8 picture area is about 50 per cent greater than that on regular 8, which means that brighter and sharper images can be recreated on the screen, or larger images can be made. Coupled with new and more efficient lamps, condenser and projection lens systems, super-8 film can now be used with classroom size groups in moderately darkened rooms.

The reels for super-8 film have a much larger spindle hole than those for regular 8. This should remind the projectionist that the film also has a different format that cannot be interchanged with other 8mm machines.

Sound on super-8 film may be either magnetic or optical. The magnetic sound is capable of somewhat higher fidelity and it can be added, erased and edited in the field. The optical track can be mass printed or duplicated with the pictures, and it cannot be accidentally erased in the field. Both are used and some projectors can be adjusted to play either.

Cartridges for 8-millimeter film enable unskilled persons, even young or handicapped children, to select and project a desired motion picture without any training and with a minimum of risk to film or machine. Most cartridges use a continuous loop so that the film is automatically rewound as it is used and ready to be repeated or

started by another. Several noncompatible cartridge systems are in use. No devices for easy and rapid inspection, winding and repair are as yet generally available.

Many 8-millimeter machines are incorporated in rear-screen viewing boxes for individual or small group use with room lights on. Due to the small size, a very bright image can be obtained. If sound is included, earphones may be used for privacy.

It must be emphasized that 8-millimeter films and machines must at present be carefully matched for sound type and position (if any), sprocket hole size and position, picture area and cartridge or reel type.

Selection

This section on selection is primarily concerned with 16-millimeter sound motion picture projectors which will project any of the large selection of films in standard film libraries. Many of the considerations also apply to 8-millimeter projectors.

The first consideration should be the degree of portability needed. If it is necessary to hand carry the machine far and often, it needs to be small, light in weight and in one case with a good handle. Serious technical compromises may have to be made for portability. It may be better to mount larger and heavier machines on wheeled carts or purchase and leave machines where they are regularly needed.

Picture quality is made up of size, brightness measured in lumens, center/corner ratio in per cent, steadiness and sharpness or resolution in the center and corners.

Image size at any distance is determined by the focal length of the lens. Two inches is standard, and many other focal lengths are available on special order. Very long or short focal lengths may have much poorer f numbers. Zoom lenses permit different degrees of magnification at higher cost.

Brightness of competitive machines is often compared by setting them side by side and observing half of each image on the same mat screen. Many precautions must be taken during this comparison. Identical films or no film should be used. A new sample of the lamp to be regularly used should be installed in each machine. Special high intensity, short life lamps should not be compared with regular lamps. The screen areas should be identical, even though different projection distances are required. A standard lens on one would not be compared with an expensive optional lens on another. This comparison is also restricted to the edges of the two images which are less bright than the centers. Both machines should have the same supply voltage by using the same outlet and cords of similar length. This assumes a circuit capable of supplying two machines simultaneously. With new and more efficient lamps and condensing systems, lamp watts should not be equated to brightness.

Xenon or arc lamps with heavy power supplies are available at added cost for much brighter auditorium projection. Special shutters that produce more illumination with more flicker may also be available.

Center to corner light ratios can be observed and compared without instruments. Somewhat more accurate comparisons can be made with an ordinary photographic light meter.

Steadiness of the image can be observed by threading an ordinary film of good quality and projecting it on a screen with the framing control turned so the frame line just appears at the top or bottom of the image. A ruler can be used against the screen to measure the amount of "jump" in fractions of an inch. The same film with the same image size should of course be used in the competitive machine. Usual "jump" is less than a half inch for an image 70 inches wide.

Resolution or sharpness of the center and corners should be made with a special test film, if available. If no test film is available a commercial film of high quality, particularly the small print credit line part, may show the capability of various lenses and machines. Many lenses have excellent center and very poor corner resolution.

Sound quality is made up of several factors considered under sound and acoustics. Sound movie projector sound requires some special considerations.

The mechanical sound produced by the film transport system in many machines is so loud that it interferes with the loudspeaker sound. In order to obtain a satisfactory signal to noise ratio, the volume must be turned up so that adjacent rooms are disturbed. One solution is to put the machine in a booth or sound absorbent box, but this is not practical in most classrooms. Since sound bounces so much from hard classroom walls, it is difficult to measure noise levels. Some comparison can be made by listening to various machines without amplifiers going in the same room one after another.

Another consideration has to do with the loudspeaker size and placement. In order to make a very portable machine, a small loudspeaker is often included in the projector case. This requires a small speaker in a poor baffle and the sound source is behind most of the audience. A separate speaker with much better characteristics and a long cord is available for most projectors. It should be placed beside and preferably half way up the screen. If only short commentary-type films are to be used in a small classroom, the built-in speaker is probably sufficient.

High-frequency response of most sound projectors is poor due to the difficulty of recording and detecting very short wavelengths on the sound track. Listening for evidence of the final s sound on speech is a common way to compare the high-frequency response of various systems.

Wow and flutter of sound projectors is best considered by using a film containing music of a piano, violin or cello. Any variation of pitch caused by uneven speed during sound takeoff is readily apparent. Common machines have wow and flutter content of less than 0.5 per cent.

Power output can be considered by noting how high a volume level can be reached before noticeable distortion is heard. Accessory loudspeakers will ordinarily improve this greatly. The difference between classroom and auditorium models is often only the loudspeaker.

Still picture projection is often wanted by teachers so that a point can be considered in detail during the film. This feature is almost impossible without harming the film or producing an unsatisfactory image. Most engineers feel that this feature should not be included in sound projectors. Teachers can obviate the need for still projection with proper preparation and follow-up activities. Filmstrips made up of stills from a movie are made to go with some films.

An adjustable aperture plate pressure control is available on some machines to vary the pressure on films of various thickness or condition. It is particularly helpful when projecting new film that is still soft or "green."

Reverse projection is often desired to repeat a portion of the film without unthreading and rethreading. If this feature is to be used, it should be checked to see that the film projects backward without loss of loops or other evidence of mishandling the film.

Silent speed is often included but seldom used at present. It should not be included in the specifications for a machine unless it will actually be used.

The length of the power cord should match the location where it will be used. Many of them are so short that an extension cord must always be used. A permanently attached or captive cord can never be borrowed or stolen. Power cords should be of the three wire self-grounding type.

Some provision for carrying the take-up reel and a spare lamp should be included with all portable machines. An accessory case may be necessary to carry the 1600-foot reel.

Automatic threading sound movie projectors have been developed to eliminate the need for some steps in threading. However, these machines still require a careful sequence of steps in order to get the machine in operation. There may also be difficulty in removing or threading a partially used reel of film. If something goes wrong with a film inside the automatic area, it may be difficult to clear the trouble. Semiautomatic machines have part of the film path, including the loops, intermittent and sound drum, opened and closed by a single lever.

If a machine is to be used in a dark auditorium, a threading lamp may be desirable.

Operation

The ordinary 16mm sound motion picture projector is a complex combination of optical, photoelectric, electronic and mechanical systems. Its operation will be considered in more detail than other machines because of its complexity and the fact that competent operation of it is the hallmark of a trained audiovisual person.

1. The projector should be placed on a stable projection stand at the distance from the screen indicated in the projection tables and opened up for operation with the lens pointed toward the screen.

2. The arms for the supply and take-up reels should be extended or attached and the film to be projected put on the upper and forward spindle, except for Victor, in which case it goes on the rear. The spindle has a round outer area and square inner area. Many film reels have a square hole at the center on one side and a round one on the other side so that they will go on only in the proper way. If the reel has two square holes, it should go on so that the film comes off in a clockwise direc-

tion from the side nearest the screen. Facing the projector, the sprocket holes should be nearest you. The reel should be locked on to the spindle with some sort of catch so it cannot fall off.

3. The take-up reel should be put on the rear or lower spindle in the same manner, except for Victor, in which case it goes on the upper front.

4. The amplifier should be turned on with the volume control low and the speaker connected if it is on a cord. If possible, the speaker should be placed beside the screen.

5. The motor and then the lamp should be turned on in a forward direction and the elevation device on the front of the machine adjusted to center the light on the screen. The projection lens should now be rotated or moved with a knob until a sharp rectangle is focused on the screen. The idea here is to get the machine in accurate position and approximate focus before threading. The lamp and next the motor are now turned off.

6. About 3 feet of film leader should be pulled off the supply reel for threading. It must not get on the floor. The leader may be labeled "head" or the title may be evident or the numbers 10, 9, 8 . . . may be seen. (The numbers are feet of 35mm film before the first scene.) "Tail" or "the end" means that the film must be rewound before use.

7. The film is attached to the upper sprocket wheel by opening a shoe or retainer and the teeth of the wheel should engage the sprocket holes. A threading line on the machine should show the correct path. The shoe should now hold the film firmly against the sprocket wheel. If there is a manual advance knob it can be rotated and the film will move slowly forward.

8. An upper loop of loose film must next be formed. Its size is indicated by a line on the machine, or room should be left for two fingers.

9. The film gate or pressure plate around the projection aperture should be opened. The lens and outside pressure plate may swing outward like a gate on hinges or there may be a way to move the outward pressure plate toward the lens a fraction of an inch. The film path through the projection and intermittent area is now open to insert the film.

10. The film should be fitted into the track provided for it past the rectangular opening where the projection light is concentrated. The claw slightly below and to the side may or may not be extended. If it is, the nearest sprocket holes may be engaged.

The upper loop must now be checked for size and the gate closed to lock the film in the projection and intermittent aperture.

11. A lower loop must be formed to divorce the intermittent motion from the sound take-off area. A manual or automatic loopsetter may be located here for easy adjustment or restoration of the loop if some damaged film failed to advance properly through the projection area.

12. A sprocket wheel and shoe may come next or rollers and stabilizers may be used. The film must follow the path drawn on the machine or the threading diagram must be carefully followed.

13. The film must now be wrapped around the hub of a flywheel called the sound drum. A roller may press the film against the sound drum or it may be tensioned by spring-loaded rollers. Most poor sound (often described as muddy) comes from improper threading around the sound drum.

14. The film next goes to another sprocket wheel and shoe which provides the pull through the sound take-off area. The sound drum is driven by the film and not by gears or belts.

15. The film next goes around some type of roller or rollers with a tension equalizer or shock absorber to prevent film breakage or sprocket hole damage from sudden tugs by the take-up reel. Additional rollers or guides may also be needed to route the film to the take-up reel.

16. The last threading operation is to attach the film to the take-up reel and wind up any slack film. The whole threading path should now be checked and if a manual advance knob is provided, it should be turned while watching the movement of film through the mechanism. Both the upper and lower loops should remain as slack film. The film should not rub against any stationary parts other than the pressure plates.

17. The motor and lamp should now be turned on in the forward direction, the silent-sound switch, if any, checked to see that it is on sound, the volume control advanced to about one third its rotation, the lens focus adjusted slightly for sharp focus on the first images, the volume and tone adjusted on the first sound, and then the frame lever or knob adjusted if the upper or lower frame line should appear on the screen. Machines with a fidelity control to focus the sound exciter system should have this adjusted for best high-frequency response. A helper can be enlisted to turn off the room lights as the screen image appears. The film and machine should be inspected regularly during projection for any possible malfunction. The most common trouble is loss of the lower loop which results in a chattering sound and a jumping or continuously moving screen image. If a loop setter is included it should be operated. Otherwise, the projector should be stopped and the film rethreaded.

18. When "The End" appears on the screen, the lamp, but not the motor should be turned off, the volume should be turned to zero or off, the room lights turned on and the motor turned off when all the film has passed through and the lamp has cooled somewhat.

19. At some convenient time the film should be rewound by routing the tail of the film back to the front of the supply reel and turning the motor to the position marked *rewind*. The lamp is not turned on during rewinding. If no position is marked *rewind*, trial and error with the motor switch may be needed. A button or lever marked *rewind* is often used to increase the speed. Another lever on the take-up reel may be operated to reduce the drag or friction during rewind.

20. The machine may now be packed up and put away. There is no danger from closing up the machine when hot.

Maintenance

The picture optical parts are cleaned much like those in still picture projectors. The lamp, reflector (if external to the lamp), condensing lenses and projection lens need regular cleaning, particularly if used in dirty locations, due to the vacuum cleaner effect of the cooling system.

The picture aperture needs particular and regular attention because the film slides through it under pressure and any dirt or foreign matter can injure the film, and it will also appear on the screen. The film gate can be opened without tools and a brush can be used to remove loose dirt. If a special aperture brush is not available, a toothbrush will suffice. If the adhesive from mending tape, film cement or soft new film (green film) has accumulated on the fixed or movable pressure plates, it must be scraped off with a wooden or plastic scraper. No metal tools must be used in this area since a slight scratch can damage every foot of every film projected. This entire area needs special, careful and regular attention to avoid film damage and to obtain good pictures. If there is any doubt about this area, a technician should be consulted.

The sound take-off area also needs special cleaning. After brushing out any dirt with a soft brush such as a clean poster paint brush, a new pipe cleaner is used to clean all the optical surfaces, including the ones usually hidden under the edge of the sound drum.

The entire film path should be checked for cleanliness and free film passage. Rollers may need a very small amount of lubrication, being very careful not to leave any where it might get on the film.

The motor and the entire mechanical system of older machines need regular oiling with the oil and instructions provided. Much damage results from too much or too little oiling. Most modern machines should be lubricated about once a year by a qualified technician in a well-equipped shop.

The take-up reel that stays with the machine should be checked to be sure that it is not pinching or damaging the film due to being dropped on a hard floor. Winding a dirty power cord on the take-up reel is not recommended.

The spare lamp that should be kept with each machine needs to be checked to be sure it is ready to use. It is very common to have the spare lamp used and the dud put in its place without any notification.

Power and speaker cords get the same rough handling as on other portable machines, and they must be checked and repaired as necessary. Lack of sound from a projector that uses a detachable speaker probably means cord or connector trouble.

Another possible cause for no sound is a burned-out exciter lamp. This small incandescent lamp can be inspected by removing a cover over it. Some exciter lamps operate only when the machine is operating. It is most often removed by pushing in and turning counterclockwise a fraction of a turn. Exciter lamps normally last much longer than projection lamps. If a replacement exciter lamp does not light, the trouble may be a defective oscillator or rectifier used to supply it with current.

Friction drive belts for operating the two reels may break or cause trouble. They can be repaired with pliers, but replacement is more often recommended. A new belt is most easily threaded into the machine by hooking it to the end of the old one. Otherwise, a new belt can usually be pushed in and through the pulley system, without disassembly, and hitched together. It might be noted that neither belt is actually needed for projection. The take-up and rewind reel can be rotated by hand for an important presentation or until a new belt can be found and installed.

Users who do not have their own technical assistance and well-equipped shop would do well to patronize a nearby audiovisual dealer who provides comprehensive service.

Assignment XII. Checked _____

Name _____

Date _____

16 MILLIMETER MOTION PICTURE PROJECTOR

1. Brand and model _____ list price _____ UL? _____

2. Power requirements: volts _____ watts _____ amps _____

3. Weight _____ dimensions _____ power cord length _____

4. Projection lamp code _____ watts _____ volts _____

5. Projection lens focal length _____ f number _____

6. At what distance will this machine just fill a 45 x 60 inch screen? _____

7. Is the loudspeaker built in, small and detachable, or large and separate? _____

   _____ cord length _____

8. Make and attach a rough but labeled sketch to show the location of each of the following parts, if they are included.

   framer   reverse switch
   still projection control manual advance
   sound-silent switch   exciter lamp
   elevation control   focus device

9. Make and attach a rough but labeled sketch to show how the film is threaded through this machine. Label particularly the upper and lower loops, the position of the picture aperture and claw, the sound drum and the shock absorber take-up idler.

10. Make another sketch or use a colored marker on the one above to show rewinding procedures and film path.

11. What loose, detachable or spare items should be carried with this machine? _____

   _____

   _____

Assignment XII.   Checked _____

Name _____

Date _____

16 MILLIMETER MOTION PICTURE PROJECTOR

1. Brand and model _____ list price _____ UL? _____

2. Power requirements: volts _____ watts _____ amps _____

3. Weight _____ dimensions _____ power cord length _____

4. Projection lamp code _____ watts _____ volts _____

5. Projection lens focal length _____ f number _____

6. At what distance will this machine just fill a 45 x 60 inch screen? _____

7. Is the loudspeaker built in, small and detachable, or large and separate? _____

   _____ cord length _____

8. Make and attach a rough but labeled sketch to show the location of each of the following parts,
   if they are included.

framer	reverse switch
still projection control	manual advance
sound-silent switch	exciter lamp
elevation control	focus device

9. Make and attach a rough but labeled sketch to show how the film is threaded through this machine. Label particularly the upper and lower loops, the position of the picture aperture and claw, the sound drum and the shock absorber take-up idler.

10. Make another sketch or use a colored marker on the one above to show rewinding procedures and film path.

11. What loose, detachable or spare items should be carried with this machine? _____

    _____

    _____

chapter 15

Single Concept Moviemakers

Background

Motion picture making has traditionally been done either by professionals in well-equipped studios or by amateurs seeking to capture family situations. With the advent of super-8 cartridge film and several versatile and nearly automatic cameras and projectors, it is possible for teachers to produce their own brief motion pictures to add just the process, situation, operation or background needed to develop a concept in the classroom. The locally produced single concept film becomes a resource to be selected and used along with 2 by 2 slides, audio tapes, overhead transparencies and discussion to produce a custom-tailored presentation for a unique class.

Super-8 film frames are smaller and less expensive than the traditional 16-millimeter format and larger than regular 8-millimeter film. Cameras and projectors based on new technology make super-8 films easy to make and project with the brightness and resolution needed for small groups (up to thirty) in well-darkened rooms (one-tenth foot candle). They are also used with smaller projectors by individuals in carrels or other study areas that may be well lighted. For this purpose they are often put in cartridges for storage, checkout and immediate use on special machines.

Although sound can be added to super-8 films, or even recorded simultaneously with the pictures, it is not now done regularly by teachers at the present state of the art, and it is not included in this discussion.

The camera is used to expose a sequence of progressively different but perfectly still pictures on a long strip of movie film 8 millimeters (nearly one-fourth inch) wide. The film probably comes in a cartridge developed by Kodak called Instamatic, although it may be produced by others too. A Japanese cartridge system called "single 8" produces the same film format for projection, but the camera cartridges are not interchangeable. When the cartridge is inserted in the camera in the only possible way the film is automatically threaded and ready for picture taking.

The camera lens focuses a picture on the film exactly as with still cameras. Since the frame is so small a lens with a focal length of about half an inch is usually used. Each frame is exposed for about 1/36 second (about the same as for still pictures) while the film is held absolutely still. After the brief exposure, a shutter revolves to block all light from the film, and a claw and intermittent mechanism advances the film exactly one whole frame, and the process is repeated eighteen times per second.

Since most films are exposed at a standard rate of eighteen frames per second (many regular-8 films were shot at sixteen frames per second), it is not possible to control exposures by varying exposure time. Instead, all exposure control must be done by regulating the diaphragm or f number of the lens, and values up to f50 may be needed for fast films on bright days.

Slow motion films are made on some cameras by exposing the film at a high rate from two to four times the normal rate. This technique apparently slows down motion when projected at the standard rate. It also consumes film rapidly and requires an adjustment in exposure.

Single frame exposure on some cameras permits the making of animated films by taking one frame at a time. It also permits studies of growth by rigidly mounting the camera and taking one picture per minute, hour or day. The single frame device is also used experimentally to produce low cost filmstrip-like materials for individual or small group study on special projectors.

The camera must have a view finder in order to aim it properly. All the better ones have the single lens reflex system described in the section on still cameras to avoid parallax and framing problems.

Most film for super-8 movies is color film of one type for daylight and indoor use. A filter is automatically positioned as the special flood lamp or a key is attached. Additional film types can be expected as the great demand for super-8 materials is satisfied. Indoor filming is no problem with modern equipment and materials.

The single concept film is used to bring some moving visual material into an educational situation. If no movement is involved, 2 by 2 slides are less expensive and more satisfactory. At least three shots are required for most single concepts. A distant or long shot is first used to establish the setting visually or to put the area to be studied into perspective. The long shot is done from a distance or with a short focal length lens. A medium shot is next used to narrow or restrict the view and lead the viewer into the material to be studied. This can be done by moving in closer or zooming in if a zoom lens is provided. The close-up shot fills the entire frame with the necessary details. In single concept film utilization, the introductory and follow-up materials are usually provided by the teacher outside of the film. Titling and labeling are sparingly used if the teacher is to be present during projection.

Very few complex or trick techniques are needed in order to make single concept educational films. Unlike longer films, these have only partial responsibility for communicating the message. A general rule is to let the subject move and not the camera. Several shots of a few seconds each from several angles or positions are better than panning and tilting the camera.

Selection

All super-8 cameras are cartridge loading so that only a door needs to be opened and the film inserted in the only way it will go. Some cameras automatically sense notches corresponding to the film speed and adjust the light-monitoring system accordingly. A window is provided in the camera so that the film brand and code can be inspected. No super-8 cameras will accommodate the old regular-8 format. The film may be transported through the mechanism by a spring motor or a battery-operated motor. Springs must be wound for each minute or two of film and batteries must be replaced regularly. A footage indicator is needed to tell how much film remains and some visual or audio indication of the end of the film is desirable.

A fine multi-element lens must be used in order to produce a very small and very sharp image on the film. Single focal length lenses around 15 millimeters are standard and generally satisfactory. Shorter focal lengths produce wide angle shots and longer focal lengths produce telephoto shots. Turrets were once popular to make it easy to rotate a variety of lenses into taking position, but zoom lenses have largely replaced them.

Zoom lenses are available with a variety of characteristics. The simplest ones have about a 2 to 1 ratio of maximum and minimum focal lengths such as 20 to 10 millimeters. This would provide a doubling of image magnification or the equivalent of moving to half the distance. More complex and expensive zooms have 5 to 1 or even greater ratios. Zooms may be operated with a lever on the lens, a knob on the camera or by a button connected to an electric motor. The motor type is referred to as power zoom.

The simplest cameras have fixed focus lenses adjusted for subjects from infinity down to 10 or 6 feet depending on the f number needed for the light conditions. These cameras are not recommended for educational work due to the need for close-up shots in sharp focus. Focusing cameras will permit minimum distances around 3 feet.

Most super-8 cameras have lenses with f numbers near 1.8 or 1.9. These very fast lenses are not expensive due to the short focal length and consequent small aperture needed. Such low f numbers permit proper exposure under low light conditions and with accessory incandescent lights.

The adjustment of aperture or f number for proper exposure can be done by hand according to general rules supplied with the film. A better way is to use a light meter adjusted for about 1/40 or 1/50 second exposure. A much more convenient way is to select a camera with a light meter directly connected to the aperture so that exposure is automatically adjusted to light conditions. Many of these meters are behind the lens so that very accurate exposures result even when panning the camera through scenes with varying light levels. When too little light is available for maximum aperture with the film used, a warning device is often visible in the view finder.

The through-the-lens or reflex view finder is very helpful in making the close-up shots that are so necessary for showing facial expressions, small parts and intricate devices.

Time lapse or animation shots require a single frame exposure capability, probably with an accessory cable release or remote control mechanism.

A lock may be available to keep the camera (on a tripod) running constantly so that the operator can move into the picture.

The completely automatic reflex camera with zoom lens is most apt to be regularly used by teachers, so its added cost is generally worthwhile.

The most needed accessory for short educational films is the light designed to match the camera. Tungsten-halogen lamps with reflectors using about 650 watts on 120 volts can be attached to most cameras. The attachment automatically changes the filter so that the same film is used indoors or out. This single light will permit movie taking up to a dozen feet or more. If automatic exposure control is provided, it will, of course, operate with the lamp. The single light can provide only flat or head-on lighting which may not be very flattering. A combination of key, background, back and fill lights will make much more pleasing but not necessarily more educational films.

A pistol grip is part of some cameras and an accessory for others. It permits steadier holding and is less tiring for extended shooting. A steady tripod is very helpful for serious and extensive shooting.

A titling kit makes it easy to compose and set up simple titles, dates, times, labels and so forth. It should be large enough so that extremely close shots are not required.

A carrying case is needed for all cameras that must be moved around to protect the mechanism from damage and dirt.

A splicer, hand-operated winders and small viewer are very helpful in selecting and rearranging lengths of film before projection. The combination is often called an editor. The splicer usually makes use of very thin mylar patches that go on both sides of the film to be joined.

Sound on super-8 film is available in several configurations that have not yet been standardized. If sound is needed, it will be necessary to purchase a complete system from one manufacturer in order to assure compatible operation.

If super-8 film is to be put into cartridges for a specific cartridge projector, then a special loading station will be needed or instructions must be given to a processing laboratory. Super-8 film from the processor can be projected on any reel to reel projector. Selecting a super-8 projector involves some of the same considerations included with other projectors.

Light on the screen is best reported in lumens from a standard ASA measuring system, but this is seldom available. Most super-8 projectors now use 150-watt lamps with large internal reflectors. Greater efficiency is obtained if the lamp operates from a transformer on low voltage. Less heat gets to the film if the reflector is dichroic so that light but not heat is reflected. Lenses with f numbers from 1.6 down to 1.2 are commonly used. The lower value results in about twice as much light with no change in the lamp. The super-8 optical system with a 150-watt lamp puts considerably more light on the screen than regular-8 projectors with 500-watt lamps, and about the same light obtained from 16mm projectors made a decade ago. With proper use of close-ups, there is no need for a very large screen image.

Projectors for this size film normally have a focal length around an inch or 25 millimeters. Other focal lengths are seldom available. Zoom projection lenses with rather limited ranges are available for slightly altering the picture size at any distance. The same result can be obtained simply by moving the projector nearer or farther away from the screen.

The standard projector speed is eighteen frames per second. Slow motion (about five frames per second), still projection and reverse projection are available on some machines. Still projection is usually discouraged due to the heat concentrated on one frame. Reverse should be used only after stopping the machine.

Some projectors will accommodate only 200-foot reels (15 minutes) and others will handle 400-foot reels (30 minutes). Reels for super-8 have a much larger center hole than regular-8 reels and they cannot be interchanged.

Most super-8 projectors have automatic threading which means that the end of the film is trimmed with a built-in device, the projector mechanism moved to "thread," the motor turned on and the film pushed in the proper opening. When the film comes out of the projection mechanism it may go directly onto the take-up reel or need a little help. The machine may go automatically into projection or need a lever moved from "thread" to "operate."

The cord length on most projectors is so short that an extension cord should probably be ordered and kept with the machine. No case needs to be ordered since all machines are built into a serviceable case. A preview or individual viewing screen may be included in the cover.

Several noncompatible cartridge projectors for super-8 are available but not considered here.

Combination reel to reel projectors for super-8 and regular-8 have been developed for individuals who have old regular-8 film and new super-8 film. These projectors are far more complex and expensive and not generally needed in education because so little regular-8 was ever used.

Operation

The purpose of the assignment at the end of this section is to prepare a brief super-8 single concept film including only one establishing shot, one medium shot and one close-up shot. The film will be processed by a commercial laboratory and later projected on a projector.

Maintenance

The only maintenance suggested for a super-8 camera consists of cleaning the lens with a soft brush and lens tissue and replacing the batteries when the meter and/or exposure control does not operate through its range and when the zoom and/or transport motors become sluggish. Batteries can, of course, also be removed and checked with a battery tester.

Assignment XIII.   Cheecked _____

Name _____

Date _____

SINGLE CONCEPT MOVIEMAKING

1. Brand and model of camera _____ list price _____

2. Film brand and code _____ cost per cartridge _____

3. Focal length of lens _____ f number_____

4. Minimum focus distance _____

5. Type of exposure control _____

6. Type of view-finder _____

7. Make and model of accessory light _____

8. Special features other than above _____

   _____

   _____

   _____

9. Concept to be taught with brief film _____

   _____

10. Distance and/or zoom setting and duration in seconds of establishing or long shot _____

    _____

11. Distance and/or zoom setting and duration in seconds of medium shot _____

    _____

12. Distance and/or zoom setting and duration in seconds of close-up shot _____

    _____

chapter 16

Film, Tape and Electrical Cord Repairs

Background

Throughout this book emphasis has been placed on understanding, selection, operation and maintenance, but not repair. This section is different. Relatively simple and often needed repairs are required for film, tape, power cords, extension cords, microphone cords and speaker cords. This work is normally done by technicians, but professional audiovisual personnel should know how to do it when emergencies arise. Broken film, tape or wires should never prevent the use of needed audiovisual instruction.

Modern motion picture film is made of a tough and durable plastic usually called acetate. It is about .005 inches thick with a binder and emulsion layer about .0001 inch on one side. It will usually burn slowly if started with a match, but it can be easily extinguished. It will go through a normal projector properly threaded hundreds of times with no noticeable deterioration. The following discussion is aimed at 16mm sound film, but it can be adapted to 35mm filmstrips and 8mm sound or silent films. If the 8mm film is in a cartridge, the film stops at the break or damage so that the two ends can be extracted for repair and then pushed back into the cartridge. Most cartridges can be opened, but special tools and a replacement clip may be needed.

Sixteen millimeter film is most often damaged by spilling off the reel during shipment (should have had end taped down or have been in a case), getting on the floor and stepped on during threading (should have pulled off only enough film for threading), reel falling off machine after threading and during machine adjustment (reel should have been locked to the spindles), mistakes in threading (should have been more carefully checked), and problems with poor film, build-up of sticky material on projector pressure plates or a defective

projector. The film may actually be broken in pieces, or, more often, sprocket holes are enlarged or torn.

A single enlarged sprocket hole will pass through all projectors since the sprockets and claw engage at least two holes each time. Two or more defective holes usually will be passed by the sprocket wheels, but not the claw. When the claw or intermittent mechanism fails to advance the film, the lower loop is immediately lost, the screen image begins to jump and a characteristic chattering sound is heard. The loop setter should be operated or the projector should be stopped and rethreaded through the gate immediately in order to avoid more sprocket hole damage. A small slip of paper can be put under the first lap of the film on the take-up reel to identify the approximate spot that needs repair during the rewind operation.

Two or three damaged sprocket holes can be trimmed carefully with scissors or a single-edged razor blade and a very thin perforated 16mm mylar pressure sensitive patch can be stuck on each side of the area. Blocks with registration pins to engage the film and patches with perforations are available for this purpose. Special ones are available for filmstrips, regular-8mm and super-8mm films.

Most often, repair consists of removing the damaged area and splicing together two ends of undamaged film. The same techniques are used when the film is broken. Film repair by removal of damaged film will shorten the length by one second for each 7 inches of film removed. A brief loss of picture is ordinarily less noticed than the loss of sound. Any extensive loss should be replaced by ordering footage from the producer and splicing it back in at a later time. Improper threading, a defective machine or negligent operation may sometimes tear the sprocket holes for the entire length of a film, and no repair is possible. Scratches for long lengths of the picture or sound track area can often be improved by commercial film treatment laboratories. Film producers often provide special trade-in allowances for extensively damaged films, particularly if they are new.

Film repair should be done with tools and materials made for the purpose. If no splicing equipment is available and the film must be used one or more times, a *temporary* splice can be made with ordinary transparent pressure sensitive tape, a ruler and a pair of scissors. The two ends of the film are each cut off square about one-sixteenth inch beyond the last good sprocket hole. A piece of tape any convenient width such as three-fourths inch is cut off square and exactly one inch long. The two good sprocket holes are exactly overlapped on a smooth surface to form one perfect sprocket hole two layers deep. One end of the tape is applied up to and centered on the double sprocket hole and rubbed down over half its length. The film and tape are next turned over and the tape wrapped tightly around the double thickness sound tracks and up to the sprocket holes on the other side and rubbed down smoothly. If properly done, the two overlapping ends of the film are held down except for the sprocket hole area, and the uninterrupted sequence of perfect holes necessary for projection is restored. It must be emphasized that this repair is *temporary*. Any tape splices

should be removed or reported to the film library after use.

No paper clips, staples, common pins, bobby pins, masking tape, first aid adhesive tape, bandaids, glue or ordinary cement should ever be used for film repair.

If a film breaks or will not pass through a machine during a presentation, it is seldom repaired on the spot. Instead, the film is unthreaded, pulled through to a good area with enough film for complete threading and the machine rethreaded. The new end of film can be placed under a complete lap of film on the take-up reel and held with one finger during the next revolution or two. During rewinding it will be necessary to anticipate this spot, or film may fly around the room!

Several makes and models of film splicing machines are available. A common machine is shown in Figure 16.2. The procedure with most machines is similar and it is diagramed in Figure 16.3. The film must be placed dull or emulsion side upward with the two ends trimmed off a frame or two beyond the last good sprocket hole. Each end of the film is clamped into separate jaws with registration pins entering the next-to-last good perforations. Cutters are provided to cut both ends of the damaged film just beyond the last good holes. The jaws holding the two film ends must now be separated to permit scraping the emulsion and binder

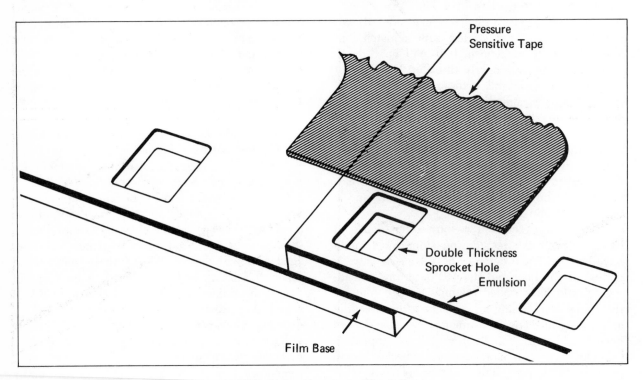

FIGURE 16.1. Temporary Film Splices

from the exposed dull sprocket hole area. A special scraper or cutter is usually provided or attached to the machine. Otherwise, a single-edged razor blade may be used for scraping. It is important to scrape off all the emulsion and not damage the film base. The emulsion is much more easily removed if it is moistened with water, but then it must be dried before cementing. Film cement is really a solvent for the plastic film base. It is applied to the approximately one-tenth inch of scraped film and sprocket hole, and then the other end of film which has no emulsion on it is quickly pressed into position over it. The film cement starts to dissolve or soften the two pieces of film and then it quickly evaporates to weld the two pieces together. The pieces should be held together with the clamps provided or hand pressure for about twenty seconds and then the jaws opened for inspection and wiping off of any excess cement. The result should be one perfect but double-thick sprocket hole that will pass through the projector with a clicking sound but no loss of loop or other problem. If the film refuses to stick together, the

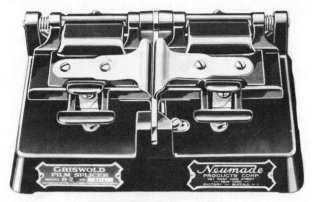

FIGURE 16.2. Film Splicer

Courtesy Neumade Corp.

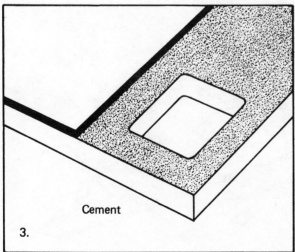

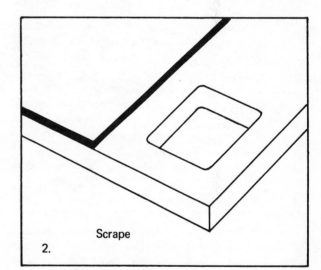

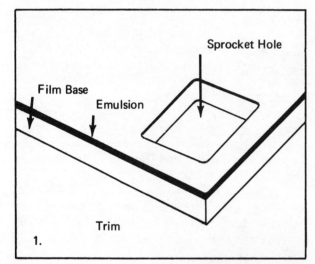

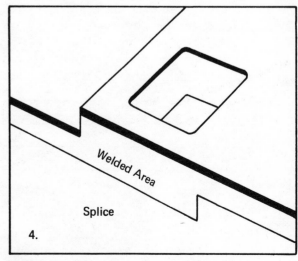

FIGURE 16.3. Film Splicing

cement may be old and need replacing. It may be that the dull or emulsion side was not scraped or scraped enough. It may be that the nonemulsion side has been waxed or lubricated so that it will need to be scraped also. Colored film is often more difficult to scrape and prepare for splicing than black and white film.

The mylar patches used for repairing torn sprocket holes can also be used for splicing film if the damaged area is removed and the two ends carefully trimmed and butted together in the splicing block. Patches are usually applied to both sides of butted rather than overlapped film.

Recording tape is somewhat easier to splice because there are no sprocket holes to be lined up and spaced. As with film, the damaged area is removed rather than repaired. Editing is also done by removing unwanted lengths of tape and splicing the wanted pieces together. Single words or even syllables from words can be removed by skilled people. Sounds on a tape can be rearranged so that sentences or paragraphs appear in different order. Awkward pauses in a speech can be shortened and often-repeated "ahs" can be removed.

Much recording tape is spliced with only a pair of scissors or a single-edged razor blade and a roll of special pressure sensitive tape, such as Scotch #41, made for the purpose. Ordinary pressure sensitive tape should not be used except in an emergency because it has excessive adhesive that bleeds out from the edges to foul up the pressure pads, recording head and capstan of the tape recorder. The special splicing tape will make permanent splices that can be used countless times.

Recording tape is spliced with the shiny side up and oxide or dull side downward, the opposite of film. The steps are diagramed in Figure 16.4. Two good ends of tape are overlapped about an

inch to be sure they are exactly lined up and then the two layers are cut simultaneously at about a 45-degree angle with a scissors or razor blade. The two ends are butted, not overlapped, and a piece of splicing tape about one inch long is put over the butt joint and rubbed down firmly with a smooth plastic or unmagnetized metal object. Next the two edges must be carefully trimmed so that absolutely no adhesive coated tape is exposed. This is ordinarily done by cutting slightly into the double thickness tapes in a smooth curve that suggests an hourglass. The resulting curve also suggests a "Gibson Girl" and the splice is often so designated. An alternative is to use a length of splicing tape only 7/32 inches wide that does not overlap the edge on either side. Metal objects used around recording tape must not be magnetized or accidental erasure may result.

Tape recording splicer blocks are often used. The block of plastic or metal has a groove that exactly fits quarter-inch recording tape. The two ends to be spliced can be overlapped and exact alignment is assured. A diagonal slot permits a single-edged razor to cut through both pieces simultaneously. The ends are then butted and ready for the tape and trimming.

More complex tape splicers are available that contain a groove similar to the blocks above, but with hold-down clamps for each side, a semiautomatic diagonal cutter and trimmer and a dispenser for the splicing tape.

A point has been made for diagonal splicing of recording tape. Two broken ends of tape can be butted together and spliced with no loss of audio

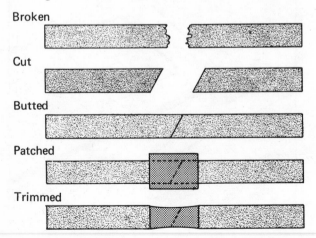

Broken

Cut

Butted

Patched

Trimmed

FIGURE 16.4. Tape Splicing

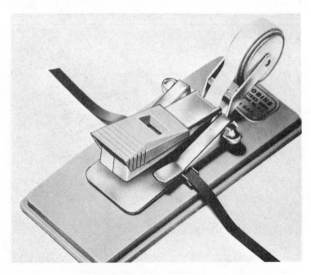

FIGURE 16.5. Tape Splicer

Courtesy Robins Corp.

whatsoever, providing the ends are not rough or distorted. The ends can be cut at right angles, but a click or pop may be heard when the tape is played. Properly prepared diagonal splices make no audible sound during playback.

A tape that has been broken and spliced should be checked in private to be sure that no ludicrous or embarassing sound has been created in the process.

A special difficulty with thin mylar tapes comes from stretching without breaking. If the resulting "stretched out" sound is annoying and not essential, the only solution is to remove it and splice the undamaged parts together.

The most common troubles with electrical cords are breakage in the middle due to being closed in a door or tripping over them, trouble where the wire enters the plug, or the connection of the wires to the posts or terminals inside the plug. A replacement cord is usually the best solution, but it may not be available when needed. Only three simple repairs will be described in this section. Obviously, all power cords should be disconnected before any work is done on them.

A power cord that has been broken in the middle can be repaired for temporary use with a pocket knife and electrical tape. The procedure is most easily illustrated with a piece of common lamp or "zip" cord. The cord is cut cleanly in two and the two wires separated about one inch on each side. One wire on each side is cut off about a half-inch shorter than the other and all four ends are stripped of insulation for about three-eighths of an inch. One long and one short end of bare wire are now twisted tightly together and taped. The other pair of wires are likewise twisted together and taped. The purpose of the staggered splice is to keep bare wires separated by both the wire's normal insulation and the tape. An extra layer of tape should be spiraled over the whole area. If the wires are colored, coded or provided with a colored thread, the same colors should be matched during splicing.

A better splice can be made with the help of simple soldering equipment. The two wires are sandpapered or scraped before twisting together. They are then twisted together and lightly coated with radio-type (not plumbing type or acid) solder flux and heated with a soldering iron or solder gun until rosin core solder from a spool will flow throughout the joint. The taping is then done as described above. This system makes a strong and safe cord repair.

Power cords often wear out where they enter the plug that goes into the wall socket. The wire is made up of many small strands to make it flexible. Through countless flexings the individual strands break, and soon the current must be carried by only a few strands which get hot. Most failures at this point can be anticipated by a hot wire and corrected before failure. The procedure is to cut the wire back a few inches and rewire it to the plug if it has screw terminals, or substitute a replacement screw terminal plug for a molded plug that cannot be repaired. The two or three wires of the power cord are colored or provided with a colored thread. It is very important with three wire systems that the wires be properly connected, which may require cutting the molded plug apart to find which wire goes to which terminal. The green wire always goes to the grounding post which is longer and usually provided with a green painted terminal. The wires for each terminal are stripped of insulation for about three-fourths of an inch, the strands are twisted tightly, preferably soldered together as outlined above, and routed around the posts and terminals and securely locked in place. A good screwdriver and pliers will be needed to form the loops around the screw terminals if solder has been used. The purpose of routing around the post before going to its screw terminal is to provide some strain relief in case the cord is pulled while the plug is still in the socket. Heavy-duty plugs also have a cable clamp to avoid strain on the terminals. A properly assembled power cord plug is shown in Figure 16.6.

White Wire and White Terminal (Neutral)

Black Wire and Brass Terminal (Hot)

Green Wire and Terminal (Ground)

FIGURE 16.6. Power Cord Wiring

Loudspeaker and microphone cables are usually soldered at both ends. The defective end can be cut off and the intact wires prepared and soldered exactly as the original was done. A crude and labeled diagram of the original connections may help make satisfactory repairs. Correct soldering technique is shown in Figure 16.7 and typical connectors are shown in Figure 16.8.

The most common error in repairing wires is to fail to get all the tiny strands properly twisted together. Soldering the strands together is often the best solution. A single misplaced strand can cause a large spark and smoke in a power cord or make a microphone or speaker cord inoperative. Most wire used in audiovisual equipment is copper

before use, or cleaning a dirty film by winding it as a soft cloth is lightly sandwiched over the moving film.

An audiovisual-type splicer costing about twenty dollars will permit professional splices, if properly used. Extra scrapers or cutters should be ordered with the machine and kept on hand so that a sharp edge is always available. Single-edge razor blades should also be stocked (never double-edge blades) for any trimming needed. Film cement should be purchased regularly in small amounts since it works much better when it is fresh. A pair of scissors should be handy for rounding off sharp corners or minor breaks and cutting out sections of damaged film.

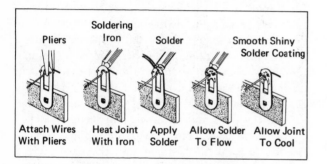

FIGURE 16.7. Soldering Technique

or copper coated with a thin layer of tin to make it easy to solder. Aluminum and most other metals are not easily soldered with common soldering equipment and materials.

Selection

The tools and equipment for maintaining and repairing film, tape and cords vary from the simplest makeshift devices through home workshop and hobby tools to audiovisual and expensive heavy-duty professional servicing equipment.

A single school or any educational unit making regular use of 16mm films should have a small service area available for use as needed. This is particularly necessary if films are to be used several times, or if films coming to the school have not come directly from a film library that inspects every film, or if films must go to another user without going through a centralized film inspection service.

A pair of film rewinders with hand cranks or variable speed motor permits the rewinding of films partially or wholly used, rapid winding forward to a part of the film desired, repairing a break in the middle of a film, inspecting a film

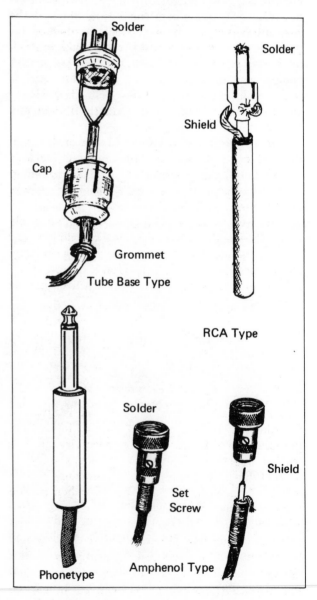

FIGURE 16.8. Audio Cable Connectors

The film service area needs spare reels for each size normally used which means 400, 600, 800, 1200 and 1600 feet. A 2000-foot reel should also be available if any films this size are ever used. The spare reels permit sending films "not rewound" to certain libraries that request it. They also permit substituting a good reel for one damaged in shipment or from dropping. A long film can also be wound through to a desired place and carried to the classroom partly on one reel and partly on another.

The film service area will also need packaging and shipping materials for borrowed films if this work is to be done there. A postal scale may save many trips to the post office. Special printed labels identify the sender, the contents (Audiovisual Materials), the requested rate (Library Book Rate) and instructions (This parcel may be opened for postal inspection if necessary). It would be well to check a proposed label with the local postmaster before printing. Special rates for these parcels should also be checked with the post office. Heavy paper, tape and twine can save much time and temper.

If many films are to be serviced, a professional film cleaning, inspecting and repairing machine should be available with a well-trained operator. These machines cost from several hundred to several thousand dollars.

Audio tapes are normally serviced with relatively inexpensive tools and equipment. Ten dollars will equip a tape service area with scissors, splicing block, single-edge razor blades, "Gibson Girl" splicer and the special pressure sensitive splicing tape.

If audio tapes are to be regularly duplicated or copied, a pair of good quality machines may be left connected together for this purpose alone. If many tapes are to be duplicated, or several copies made at a time, a professional tape duplicator should be purchased.

A bulk eraser for audio tape will remove all recorded material before tapes are used for another purpose. This is particularly necessary if one teacher might criticize another teacher's work, or if confidence (i.e., guidance-counseling) is involved. Total erasure is also needed when several track configurations may be used with the same tape.

A drawer or box of tools, Figure 16.9, should be assembled for electrical cord repairs if this job is to be done. It should contain regular and Phillips head (star point) screwdrivers of various smaller sizes, pliers (5- or 6-inch size), diagonal cutters (5- or 6-inch size), black plastic electrical tape, a knife (pocket jack knife or small paring knife), scissors, a soldering iron such as a 100-watt soldering gun, solder paste (radio type) and solder (rosin core). A small vise is often very helpful for holding parts while working on them. A small electrical tester such as a volt-ohm-milliammeter is also helpful for checking the presence of voltage (150 VAC scale) at a particular point or for checking circuit continuity (low ohms scale) with the wires disconnected.

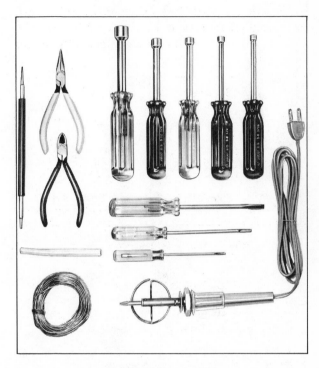

FIGURE 16.9. Tool Kit

Courtesy Heath Co.

Operation

The purpose of the laboratory assignment is to practice the repair of film, tape and wiring using the techniques described and instructions included with any machines used.

Maintenance

Maintenance of the tools and equipment for a local repair area is not very complex. Cutting edges should be replaced as necessary or possibly sharpened. Film splicers can be cleaned by using new film cement to soften old and accumulated cement. Scraps of audio tape and splicing tape must be removed from the tape splicer. Replacement cutters are available for the better tape splicers.

Assignment XIV. Checked _____

Name _____

Date _____

FILM, TAPE AND ELECTRICAL CORD REPAIRS

1. Staple or tape below a sample of 16mm film carefully spliced by the *temporary* pressure sensitive tape and scissors method. Do not use audio-splicing tape for this purpose.

2. Film splicing machine brand and model _____

3. Film splicing cement used _____

4. Sample of film splicing done on this machine

5. Sample of tape splicing done only with scissors or razor blade and splicing tape.

6. Tape splicing machine brand and model _____

7. Splicing tape brand designation and width _____

8. Sample of tape splicing done on this machine

9. Prepare an electrical cord end and attach a replacement type 120-volt plug. It should be checked by the lab instructor before use.

10. Prepare and solder wires or wire and a connector as provided by the lab instructor.

chapter 17

Closed Circuit Television

Background

Closed circuit television can mean anything from a simple camera and a home-type receiver used to magnify demonstration materials in a small classroom to very complex state-wide systems for delivering a major part of the presentations for a curriculum.

This discussion is limited to relatively simple, self-contained and portable systems that enable a moderately skilled person to capture the sights and sounds of an educational presentation, record them on television tape and reproduce them on one or more monitors or television receivers in the same or a nearby location. Studio, system-wide and broadcast television are not considered since they require specialized professional and technical personnel.

The basic system for capturing, storing and reproducing moving pictures and sound electronically is rather complex and not understood by many of the educators who now must select and operate it.

The television camera, Figure 17.1, converts a moving picture into an electronic signal which is carried over a wire and converted into a magnetic pattern on a tape similar to but wider than audio tape, then converted from the magnetic pattern back to an electronic signal and delivered over a wire to the monitor or receiver which recreates the original pictures and sounds. If delayed reproduction is not needed, the camera may feed the monitor directly.

The television camera has a lens similar to the ones used in motion picture cameras. Many lenses are used both for 16mm movie cameras and closed circuit television cameras. The television camera lens focuses the scene to be captured on a light sensitive plate behind a transparent window in a vacuum tube called a vidicon, which is about an inch in diameter and 5 inches long. The image within the tube is about the same size as the image on 16mm motion picture film.

It is impossible to capture electronically all the picture detail at once as is done with a photographic camera and film. Instead, a scanning system is used so that only the light or shade of one small picture area or element at a time is sensed and converted to electricity. The scanning system is similar to reading a page of printed material. The eye and brain cannot handle the information on a page in one block, but it can easily do so by scanning the information along one line at a time.

Picture information in the form of electric charges on the plate within the vidicon tube is scanned by a very sharp pencil or beam of electrons formed by a cathode and plate or anode with a pinhole in it. The beam of electrons can be focused and moved over the entire picture area in a very precise scanning pattern by means of signals developed inside or outside the camera. As the electron beam strikes the electrically charged picture areas, it will be attracted or repelled depending on the charges present, and a minute variation in electrical current can be detected that is directly related to the picture information at each point. This minute signal is amplified by tubes or transistors within the camera and sent over a wire to the recorder or monitor.

Scanning is done in North America, but not the rest of the world, in 525 horizontal lines beginning at the top left corner (as with a page of print) and moving across the image from left to right to detect the shades of gray (including black and white) in one narrow band of the picture. The beam is then very rapidly returned from the right end of line one to the left end of line three which is then scanned. When all the 262.5 odd-numbered lines have been scanned, the beam is very rapidly moved from the lower right corner to the top cen-

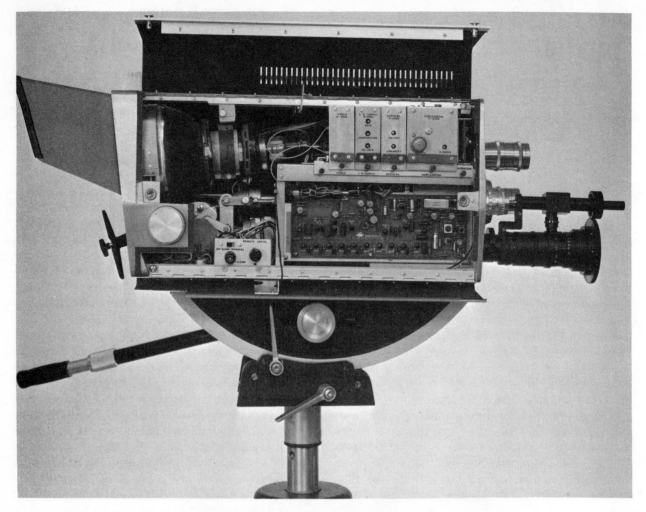

FIGURE 17.1.   Television Camera (Covers open)

Courtesy Diamond Power

ter where it begins to trace the even-numbered lines of the picture until it has completed 525 complete lines. The scanning process is shown in Figure 17.2. The alternate scanning of odd-numbered and the even-numbered lines in the picture is called interlaced scanning, and the purpose is to reduce flicker on the screen of the receiver or monitor.

The timing and positioning of the scanning beam of electrons must be very accurately done in order to produce high-quality pictures and a signal that is compatible with various recorders and monitors. The 525 lines to produce one complete picture or frame (same meaning as with movies) must be traced in exactly one-thirtieth a second to produce a frame repetition rate of thirty still pictures per second. The 262.5 alternate lines which constitute one field are traced in one-sixtieth a second to produce a field frequency of 60 per second in order to reduce flicker. Each line is scanned in 63.5 micro-

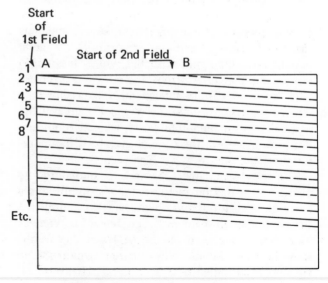

FIGURE 17.2.   Television Scanning Process

seconds, or at a frequency of 15,750 lines per second.

Not only must the scanning beam trace the picture area very accurately, it must also be blanked or extinguished during its very rapid retrace from right to left and from bottom to top. The pulses required to do this must also be included with the picture information and recorded or delivered to the monitor.

Even with the complex and precise system described, only a relatively crude image is possible compared to photographic recording and reproduction.

All television images in North America, with the exception of special nonbroadcast and expensive high definition systems, have only 525 horizontal lines of information, no matter how big the screen or how good the system. Each horizontal line can have about 600 bits of shading information along its length in the best closed circuit systems, but only about 350 lines of resolution are possible with broadcast or ordinary closed circuit equipment. A typical television picture contains about 150,000 picture elements which can be changed thirty times per second.

The synchronizing pulses for timing and locating the scanning lines may be included in the camera or in a separate unit. A separate and common sync generator is advisable for multiple camera and critical work.

The picture and sync information produced at the camera can be carried over a special wire within a shield, called a coaxial cable, directly to a television monitor which may look like a home receiver. The sync pulses are removed from the signal inside the monitor and used to move an electron beam within the evacuated picture tube in exact synchronism with the vidicon beam. The electron beam in the picture tube strikes a fluorescent coating on the inside of the face and produces an amount of light dependent on its strength. The 525 lines, called a raster, can be easily seen when close to the screen. The picture or video information is used to modulate the intensity of the electron beam as it scans the picture tube and recreates an image. The combination of picture and sync information together is called composite video, and it can be sent over a coaxial cable for a thousand feet, or more with special amplifiers. Sound cannot be sent on the cable with composite video, so a separate audio line (pair of wires) must be used. Monitors usually have audio amplifiers and speakers for reproducing the sound. If not, then a separate sound system must be used.

In order to transmit several pictures and accompanying sound signals to one or more standard television receivers, the picture and sound are used to modulate radio frequency or RF oscillators on the standard VHF-TV channels, 2 through 13. These signals are very much like commercial broadcast television signals, and they can be distributed throughout a building distribution system simultaneously with broadcast signals. The RF system is very flexible and inexpensive receivers can be used, but the picture and sound quality are about half of that obtainable with composite video and separate sound systems.

All television images have a ratio of picture height to width or aspect ratio of three to four. The picture is always 3 units high and 4 units wide. This imposes restrictions on visuals used on television. All motion pictures, filmstrips and horizontal oriented overhead transparencies and 35mm transparencies have approximately this same 3 to 4 aspect ratio.

Television tape recorders are like audio tape recorders in that a complex electrical signal is used to magnetize particles on a tape that is pulled past a recording head at constant speed. All television recorders also have one or more audio tracks for separately recording the sound exactly as in audio tape recorders. The recorder tape can be rewound and immediately passed over the heads again to reproduce the electrical signals which can be fed to a monitor and loudspeaker or into an RF modulator and to one or more receivers.

Television recording is much more difficult than audio recording due to the tremendously greater amount of information that must be handled. The best audio recorders will handle frequencies up to about 15,000 cycles per second. The recording and reproduction of ordinary home television quality pictures require frequencies to about 4,000,000 cycles per second (4 megahertz) and high-quality pictures require about 8 megahertz of bandwidth. Frequency response in tape recorders is increased by narrowing the magnetic gap where the recording is made and increasing the motion between the tape and recording head. With the narrowest practical gaps it is still necessary to have tape to head velocities of about 1000 inches per second. Such velocities can be attained with very long lengths of narrow tape, but such a system is not now practical. All common television recorders now make use of wide tape and a rotating head or heads that record a series of narrow bands diagonally across the tape.

All broadcast television is done with a four head system called quadrature on 2-inch wide tape with precise conditions so that any tape can be played on any machine at any broadcast station. This is a completely compatible but very expensive system.

No such compatibility exists with educational or closed circuit television tape recorders. The tape may be one-half, one or two inches wide. The tape may move from less than 4 up to 12 inches per second. The video and audio tracks may be arranged in any of several formats. It is essential that a television tape be played on exactly the same make and model of machine that was used for recording it. A typical television tape and magnetic pattern is shown in Figure 17.3.

Educational television tapes can now be exchanged only among people who use the same machines, or by going through a duplication or "dubbing" center which has a variety of machines that may be interconnected as needed. Such dubbing or copying inevitably results in lower picture quality.

Television tape is made of very thin mylar or polyester plastic with the oxide coating securely and carefully applied. During recording and playback it is necessary to have the head pushed into the tape for good contact, and the resulting deformation of the tape is called tenting. Only the tape recommended by the recorder manufacturer or a reputable television engineer should be used to minimize noise or dropout in the picture, flaking or rubbing off of the coating and excessive wear of the heads or heads. Neither tape nor heads have anything like the long life of comparable audio equipment.

Selection

A closed circuit television system consisting of a camera, camera support, lights, sound equipment,

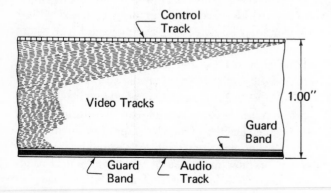

FIGURE 17.3. Magnetic Pattern on Television Tape

recorder, monitor, accessories and cables should probably be purchased as a package from a nearby competent and accommodating dealer. Many television systems have been purchased, that for one reason or another are doing no significant educational job.

Cameras differ widely in their capabilities and characteristics. They should be UL approved and provided with door interlocks that disconnect the power if any harm could result from opening them.

The dimensions and weight of the camera may be very important if portability is needed, and unimportant otherwise.

A view finder or monitor mounted on the camera is very helpful for an operator who is remote from the audience or recorder monitor. All the better and more complex systems, except remote control systems, use view finder cameras. The size of view finders is normally the diagonal measurement of the picture in inches. If more than one camera is to be used, a red tally light should be visible to the operator and talent to indicate which camera is in use. A view finder hood is desirable for checking the picture under bright overhead lights.

The camera lens feeds the light and picture information into the system. A "C" mount or connector is most common, and recommended. A low f number such as 1.9 is needed in order to collect enough light for operation in a classroom with about 30 footcandles of illumination. The most common focal length lens is about one inch or 25 millimeters, which has an angular view of 30 degrees or a width of about 6 feet at a distance of 12 feet. A shorter focal length such as 12.5mm would provide the same coverage at half the distance. A telephoto lens with a focal length of 100mm would be useful for enlarging small objects without the necessity for moving in close. Several fixed focus lenses can be mounted on a turret fixed to the camera. Turret operation should be from the back or operating position of view finder cameras. Instead of a variety of focal lengths, zoom lenses are becoming very popular but at greater cost than several fixed focus lenses. Zoom lenses are available with various ratios between minimum and maximum focal lengths, various focal lengths, various f numbers and manual or motor operated. A 4 to 1, 13mm to 52mm, f2 zoom lens might be typical. Lenses for extreme close-up work will need one or more extension tubes.

There are various qualities of vidicon pick-up tubes from industrial types to broadcast types. The quality of the picture needed must be compared with vidicon tube cost.

The overall picture quality produced by the camera is best described in terms of horizontal resolution or number of distinct lines that can be distinguished across the face of a high-quality monitor. Industrial television cameras can be expected to produce 200 to 400 lines. Inexpensive school equipment for noncritical purposes should produce 400 to 650 lines, and high-quality closed circuit cameras should produce 650 to 800 lines. It might be debated that the recorder, distribution equipment and receivers will only produce 350 lines, but that 350 lines will result only if the camera produces many more lines. A test pattern for judging image quality is shown in Figure 17.4.

Most newer cameras contain transistors on printed circuit boards for compact construction and rapid warm-up. Tube equipment can do an equally good job if size and warm-up time are not important. The transistor equipment may require module replacement rather than repair if trouble develops.

The sync and interlace circuits are built into the simplest cameras and located outside the better ones. The interlace may be random, or it may be precisely controlled by crystals. High-quality pulses can be assured, at a price, by requiring EIA (Electronics Industries Association) standard sync. A common sync generator is needed to drive multiple cameras with switching and no momentary loss of picture (roll over) or to make use of split screen images through special effect generators. Simple random interlace circuits will suffice for single camera noncritical uses.

To prevent burning a spot or line in the vidicon photosensitive plate if the deflection circuits should fail, a scan-loss protection circuit is available with some cameras.

Sweep reversal is available on some cameras to turn the picture upside down or reverse it sideways. Polarity reversal makes black on white lettering appear to be white on black. These effects are most needed with slides or graphic visuals used to illustrate presentations.

The camera may have composite video output to feed a monitor directly, or RF output on one or one of several VHF-TV channels to feed one

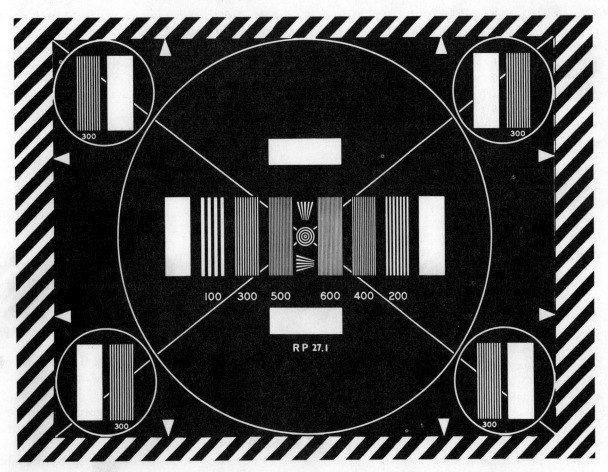

FIGURE 17.4. Television Test Pattern (SMPTE)

or more television receivers. Tape recorders require composite video. Receivers require an RF channel which may come from the camera or an external modulator. Monitors will operate on composite video and may also have a tuner for RF channels.

If the camera and its associated equipment are to be operated outside the same room, cable lengths and possibly additional equipment may need to be considered.

Cameras need to be properly adjusted for optimum picture quality. The ease, rapidity and accuracy of the setup procedure should be considered. After a few minutes of warm-up, the picture should remain stable for continued operation.

A substantial camera support should be selected to match the camera weight and the amount of maneuverability required. The most common portable support consists of a friction head, elevator column, tripod and dolly. This combination permits easy and smooth panning, tilting, locking in any position, raising and lowering of the whole camera and wheeling it about the floor. Pedestals are used for larger and nonportable cameras. Cameras are also put on posts attached to tables, stands and wheeled carts for special purposes. Some package units on wheels hold camera, recorder, monitor and accessories. They may be difficult to transport up and down stairs or in and out of buildings.

Additional lighting will make the greatest improvement in closed circuit television at the least cost. There have been far too many demonstrations of minimum television pictures with minimum light. A general level of approximately 100 footcandles will enable the aperture (f number) of the lens to be adjusted so that sharper pictures and much less critical focusing result. Additional lights as in photography may be used for emphasis and more pleasing pictures. Several portable lighting packages using small tungsten-halogen lamps and reflectors are available. Silicon-controlled rectifiers can be used to dim these lamps. Photographic floodlights will also work well. Close-up work or demonstration magnification on a table may require only a pair of desk or gooseneck incandescent lamps. The color temperature of lamps for monochrome vidicon television is not critical.

Sound for simple closed circuit work is normally picked up with a lavalier microphone around the neck of the teacher. A high-quality, nondirectional low impedance dynamic microphone is usually recommended. The microphone feeds directly into the recorder or to the monitor if a recording is not made. An audio signal from a record player or other source may also be fed to the recorder or

monitor, but if mixing of music and commentary is wanted an accessory mixer may be needed. A wireless microphone using a small radio transmitter and receiver will free the teacher from wires but at more cost and complexity. If a receiver is to be fed instead of a monitor, the microphone is connected to the modulator simultaneously with the video, or a separate audio amplifier may be used.

If several cameras are used or if there is some distance between the camera and the place where the signal is used, an intercom system may be needed. Such a system may be included in the camera or an accessory. The combined earphone-microphone headset is most useful. One earphone may reproduce the program and the other one reproduce the interphone instructions.

Choosing a slant track or helical scan television tape recorder for closed circuit use is very difficult due to the variety of incompatible makes and models. Most of them also leave much to be desired in picture quality, sound quality, ease of operation and reliability. Unless high level technical assistance is available on the staff, only a recorder sold and serviced by a nearby dealer should be considered.

If tapes are to be exchanged with another user, it is essential that the same make and compatible model of machine be purchased.

Line resolution is the best commonly available indication of picture quality. The least expensive machines provide only about 150 lines and the best around 400 lines. Head speed, number of heads, tape speed, tape width, writing speed and frequency response are all important, but only to provide improved line resolution.

The picture should also be essentially linear, completely stable and free from noticeable distortion and noise (extraneous light flashes).

The weight and size of the recorder may be important for portability. Common recorders weigh up to 150 pounds. Handles will be needed for portability, but the presence of handles does not mean that an ordinary person can easily transport it.

Rewind time may be important if long tapes are to be replayed while a group is waiting.

More than one sound track may be desirable in order to add comments or a substitute sound track at a later time without destroying the original sound.

Slow motion or stop motion may be desirable for activity or motion analysis and study. Stop motion may unduly wear and weaken the tape at the one scan line diagonally across the tape.

Remote push-button operation from a distance may be desirable, particularly if one person must handle several operations.

All recorders have an automatic erase system for both audio and video before a new recording is made. If two audio tracks are provided it may be possible to erase and rerecord one without disturbing the other.

Video record-playback heads have a limited life up to several hundred hours and must then be replaced. The guaranteed life, cost and ease of replacement are important considerations.

The maximum recording time on one tape reel varies considerably. So does tape cost and cost per hour.

A complete sound playback amplifier and loudspeaker may be included in the recorder or it may be available as an accessory. An alternative arrangement is to have the audio playback facilities in the monitor.

An editor of some kind may be included or available as an accessory to permit locating a particular place on the tape and then substituting another picture and sound segment. Cutting and splicing video tape is not recommended, although it is commonly done. As the spliced tape goes through the machine, the picture usually becomes very unstable and the splicing tape may interfere with the mechanism.

Meters are necessary in order to adjust picture and sound level during recording and possibly for adjusting playback and other circuits for best results.

Battery-operated portable recorders for limited quality pictures and sound are becoming available. The batteries can be recharged overnight.

The recorders considered here are for monochrome television, which is standard in education at the present time. Many of the newer models have provision for adding modules so they can record and play back color signals. This feature probably indicates that existing features are high in quality because color requires attention to all details, but the color cameras and monitors for use of the color feature are as yet very expensive and complex.

A monitor or receiver is necessary to view the picture. It may be a very high-quality studio monitor or a typical home television receiver. A monitor ordinarily accepts only composite video signals and audio from a recorder or microphone. It may also have a tuner for RF channels. A receiver will not accept composite video unless a technician has modified or "jeeped" the circuit inside the set. Monitors normally can provide much better pictures and sound than ordinary receivers.

FIGURE 17.5. Television Monitor—Receiver

Courtesy of RCA

The size of the screen is directly related to the number of viewers to be served. A 23 inch screen is most often used for classrooms and it has a screen that measures 23 inches across the diagonal, not the width. Such a receiver should have the farthest viewer no more than about 20 feet distant. Much smaller monitors or receivers can provide equally satisfactory pictures for small groups or an individual. There is actually as much information in the small pictures as the large one.

Many receivers have small and inadequate amplifiers and speakers for classroom groups. In addition, some speakers are not front mounted so that all sounds must bounce off a wall before they are heard. It is important to have good amplifiers and front-mounted speakers so that every word can be heard. More of the message is usually carried by the audio than the video.

The special receiver-monitor sets made by several manufacturers are usually much better than home receivers for classroom viewing. Most of these have several inputs and outputs for connection in a variety of ways.

Monitors or receivers for group viewing need to be placed on stands about 44 inches high with well-spaced legs for stability.

Home television and poor quality or poorly adjusted closed circuit monitors have accustomed many people to much poorer pictures than those possible at the present state of the art. Adjustments can be carefully made to improve many pictures. The raster should just fill the height of the picture tube by adjusting the vertical height control. It should just fill both sides of the tube by adjusting the centering and width controls. Many receivers overscan the tube so that as much as half of the picture information is lost. An object at any place on the screen should be the same size, or a circle (usually a test pattern or station identification slide) should appear round by adjusting the linearity controls. The lines making up the picture should be bright enough for easy visibility in a partially darkened room if a large group is involved, or in a fully lighted room for a small group by adjusting the brightness control. About eight shades of gray between white and black should be visible by adjusting the contrast control. The picture should be absolutely stable vertically by adjusting the vertical hold control and horizontally by adjusting the horizontal hold control. The picture should be free of extraneous black or white flashes (noise) by adjusting the fine tuning control, assuming a good signal coming into the set. Some of these controls are exposed on the front of the set. Others may be behind a door or available on the back or sides. Not all the controls described may be included.

It is often desirable to add motion and still pictures from projectors to a closed circuit system. In a complete system, this is done with a special fixed camera and a multiplex device that enables several projectors to focus directly into the camera through mirrors or prisms. With the simple systems considered here, it is more often done by focusing both the projector and the camera on the same screen and adjusting screen brightness by moving the projector back and forth until a satisfactory picture is obtained. A translucent rear projection screen puts the projector out of sight and often produces better pictures. The images can be reversed with a mirror or with a sweep reversal switch, if one is available.

The cables for all the pieces need to be long enough for the setup desired. It may be necessary to order special lengths of cable with special fittings. Making up special cables in the field can be very time consuming, and they may be unreliable.

Operation

The great variety, complexity and lack of standardization in the new and rapidly growing area of portable closed circuit television equipment makes general operating instructions very difficult. Instead, it is suggested that the operating instructions provided with the equipment be studied carefully and followed to the letter. Indiscriminate knob twisting with the hope that good pictures will magically appear is definitely not recommended. There is also the real danger that the vidicon camera tube will be damaged.

Special precaution number one: The camera should not be pointed toward the sun or any bright light, or the vidicon may be damaged. Do not leave the camera pointing toward any brightly lighted scene for more than a minute without moving it at least slightly, or an image may "burn in" and remain as a ghost later on. Cap the lens whenever the camera is not in use.

Special precaution number two: The camera if on a friction head must be unlocked from its cradle or head by loosening hand-operated locking devices and then held firmly during use. While still holding the camera, the locking devices must be tightened when use is completed. Many cameras have been severely damaged by an operator simply letting go of an unlocked camera.

An operating sheet at the end of this section is available for reporting the characteristics of the camera, recorder and monitor-receiver comprising a closed circuit television system.

Maintenance

Only the maintenance specifically recommended and described in the manual accompanying this equipment should be attempted by ordinary operators in the field. Any other work should be done by a qualified resident engineer or a person provided by the service department of the dealer who supplied the equipment.

Magnetic recording tape should be kept absolutely clean at all times. The machine should be kept covered when not in use and the entire tape path cleaned regularly. Smoking should not be permitted in the taping area. The entire taping area should be vacuum cleaned regularly.

Tape should be stored on the reels provided. The loose end of the tape should be stuck down with a small piece of pressure sensitive tape. The reels should be handled by their hubs and not their flanges in order to avoid pinching the tape or bending the flanges. All reels of tape should be stored in cartons in an upright position. If the tape on the reel does not appear to be uniformly wound, an adjustment of machine tension is probably needed.

Tape is often damaged along an edge by improper tape guide adjustments in the machine or by a damaged reel. This may also happen if some strands on the reel protrude beyond the usual edge due to scattered winding.

The temperatures and humidity for human comfort are also best for tape storage. This means 60 to 80 degrees and 40 to 60 per cent. If tape has been exposed to extreme weather during transport, it should be allowed to attain room temperature gradually before use.

Tape can be accidentally erased but only by being very close to very high magnetic levels. It should not be allowed within a few inches of electric motors, transformers or loudspeaker magnets. Under ordinary conditions magnetic tape will hold its message indefinitely.

Assignment XV. Checked _____

Name _____

Date _____

CLOSED CIRCUIT TELEVISION

A. CAMERA

    1. Brand and model _____ list price _____

    2. Power requirements _____ UL approval? _____

    3. Dimensions _____ weight _____

    4. View finder? (size) _____ tally lights? _____

        Hood? _____

    5. Mounting: head _____ elevator _____

        Tripod _____ dolly _____

    6. Lens or lenses available (focal lengths and f#) _____

        _____

    7. Sweep or polarity reversal? _____

    8. Composite video output? _____ RF output (channel #) _____

        _____

    9. Intercom system? _____

  10. Advertised resolution of camera _____

  11. Type of sync provided _____

  12. Accessories used with camera _____

B. RECORDER

    1. Brand and model _____ list price _____

    2. Power requirements _____ UL approval? _____

    3. Dimensions _____ weight _____

    4. Tape width _____ tape speed _____

    5. Advertised resolution _____ maximum recording time _____

6. List the video inputs provided _____

_____

7. List the audio inputs provided _____

_____

8. List the video outputs provided _____

_____

9. List the audio outputs provided _____

_____

10. How is proper video recording level indicated? _____

11. How is proper audio recording level indicated? _____

C. MONITOR OR RECEIVER

1. Brand and model _____ list price _____

2. Power requirements _____ UL approval? _____

3. Dimensions _____ weight _____

4. Antenna connections provided: VHF _____ UHF _____

5. List the video inputs provided _____

_____

6. List the audio inputs provided _____

_____

7. Screen diagonal in inches _____

8. Size and location of speaker _____

9. Controls exposed on front of set _____

_____

10. Controls behind door or otherwise available for adjustment _____

_____

_____

chapter 18

Mounting and Laminating Machines

Background

Flat paper printed materials are readily available to supplement and illustrate practically every lesson that is taught. The small size of these materials and the detail included makes it difficult for a group to observe together, so they are commonly passed around a class, left for individual perusal on a study table, posted on a tackboard, assembled in a loose leaf book, or put in a vertical file for checkout as needed.

Ordinary flat paper printed materials may have just the content needed for a lesson, but they do not have the durability needed for repeated handling and long-term use. They may also have distracting material on the reverse side. They are very apt to be lost or appropriated for personal use if they appear to be ordinary scraps or sheets of paper. Some method for mounting or laminating these materials is essential for more than one-time or incidental use.

The simplest method for use of clippings or tear sheets from newspapers, magazines, advertising materials and so forth is to slip them into transparent plastic envelopes that are designed for salesmen to use in their visits to customers. These are available in various sizes and arrangements, but most often for 8 1/2- by 11-inch materials and for an ordinary loose leaf notebook. Both sides of a page can be read. A plain sheet of paper or cardboard can be inserted to hide unwanted material. A folder or notebook can be made up for any number of uses. The new plastic called mylar or polyester is somewhat more expensive than acetate or ordinary plastic, but much tougher, particularly if holes are punched for loose leaf notebooks.

Another simple mounting method makes use of the ubiquitous transparent pressure sensitive tape and a piece of stiff paper or cardboard. The tear sheet is accurately trimmed on a paper cutter and positioned on a larger mounting sheet with a weight to hold it in place. Strips of tape are then neatly applied along the four edges. A single-edged razor, X-acto or frisket knife can be used to cut the tape at the corners for a very neat job. Transparent mat finish mending tape will be nearly invisible when applied in this manner. Pressure sensitive tape is also available in many colors, both transparent and opaque, and in many widths, for mounting sheets with a border accent.

Double-coated pressure sensitive tapes and various special mounting tabs are also available for mounting sheets.

The double-coating rubber cement method requires no equipment and produces excellent results. The sheet to be mounted is carefully trimmed and positioned over a slightly larger piece of cardboard or heavy paper and the corners are lightly indicated with a soft pencil. The tear sheet is then turned face down over scrap paper and a thin coating of rubber cement applied evenly to the back with the brush provided in the bottle, making sure to go slightly beyond the tear sheet area in all directions. The backing sheet is now coated with rubber cement out to and slightly beyond the area to be covered by the tear sheet. Both coatings are allowed to dry for several minutes so they are tacky but not sticky. Two ordinary sheets of household wax paper are next overlapped an inch in the center of the mounting on top of the cement. The tear sheet is then carefully positioned over the wax paper, observing the corner pencil markings. First one half and then the other half of the wax paper can be removed while holding the tear sheet exactly in position. The sheet is then smoothed down and the excess cement around the edges is rubbed off with the fingers. If positioning is not critical, it can be done without the wax paper. The adhesion, however, is so great that no repositioning, even of a small part, can be done without destroying the paper. Rubber cement is also available at higher cost in pressure cans for spraying instead of brushing.

Dry mounting has earned an excellent reputation in audiovisual education because it is so easy, permanent and rapid. It is ordinarily done with a special dry-mounting press and sheets of dry-mounting tissue that are applied between the tear sheet and the mount. The dry-mounting tissue is dry and in no way sticky when it is at room temperature. At a temperature near the boiling point of water, 200 degrees to 225 degrees F, the tissue melts and penetrates both the tear sheet and mount. When allowed to cool the tissue again hardens to form a permanent bond. A small tacking iron is usually used for arranging the sandwich before putting it in the press.

Dry mounting is accomplished by tacking an oversize piece of dry mount tissue to the back of the sheet to be mounted by touching the hot tacking iron near the center. The tear sheet and tissue are next trimmed together on a paper cutter to the exact size wanted. The two sheets are then put over the mount and tacked in position at one or two corners by lifting the tear sheet to expose the tissue. A thin clean sheet of paper is then put over the tear sheet for protection and the whole combination put in the preheated press (about 225° F). Heat and pressure are applied for about fifteen seconds and then the fused materials are removed and allowed to cool before handling. Weights may be used to keep materials flat during cooling.

If bubbles appear or incomplete fusion occurs, the likely cause is moisture in the tear sheet or mount. In moist climates, it is wise to put both tear sheet and mount in the press separately for a few seconds to dry them thoroughly before starting the mounting process. Thin protective sheets of disposable paper are usually used over all clean materials in case there is dirt on the hot plate of the press. If bubbles or incomplete fusion is evident, a second heat and pressure application will usually be effective. Heavy materials may require longer processing times.

If a dry mount press is not available, the same materials can be used with an ordinary dry flatiron on a smooth surface covered with several sheets of paper. The iron should be just hot enough to "sizzle" when touched with a moistened finger. The tear sheet should, of course, be covered with a thin protective sheet and the iron should be kept in motion working from the center outward.

The dry mount system hides one side of the paper permanently. If both sides are important, then two copies of the material must be obtained. Any such materials should be collected in pairs when the magazine or newspaper is current.

Dry mounting cloth with thermoplastic material on one side is also available for backing fragile, old, or much used materials. It has often been used to back maps. It can be used in the dry mounting press or ironed on with a flatiron.

Lamination means that a transparent adhering plastic covering is applied over the sheet to be preserved. Ordinary transparent or mat finish pressure sensitive tape can be used for this purpose, but several widths would ordinarily be needed. Sheets of self-sealing acetate or mylar are available on a glassine or nonsticking paper for use with no equipment other than scissors or a paper cutter. The glassine paper is removed from an oversize sheet of the plastic with the nondrying adhesive facing upward. The tear sheet is smoothed face down over the adhesive. The exposed border of adhesive can be used to attach the tear sheet to a mount, or a second sheet of the self-sealing acetate can be used to cover the other side and make a transparent sealed edge all around that can be trimmed to a desired size. It may be necessary to apply some pressure to smooth out all areas. Since the adhesive ordinarily does not dry out inside the "sandwich," some care must be taken in storing and using these laminated materials.

Several laminating films are available with a very thin coating of transparent thermoplastic material which is not sticky when cold but very sticky when hot. These materials must be used with heat, but they result in higher quality and permanently laminated materials.

A special laminating film is available for infrared or heat process office copying or transparency making machines. Special arrangements are needed in order to obtain enough pressure at the instant the heat is applied inside the machine. This process is not now recommended.

Another laminating film is available for dry mounting presses. It comes in sheets that are folded over the material to be laminated with the duller side facing the printed surfaces. A thin protective sheet of paper should be placed over each side, and a piece of heavy cardboard or plywood must be put under the "sandwich" in order to increase the pressure during the fusing process. The temperature should be set to 270 degrees and timed for about fifteen seconds, or longer if adhesion is not complete. This material can also be used with a dry and hot flatiron, but some skill and practice may be needed.

The best quality laminating is done in a machine made especially for this purpose. It contains rolls of thin mylar coated with dry transparent thermoplastic material. Two heated metal rollers apply, fuse and firmly press the film to any paper or thin card stock that is fed into the machine. Any length of paper up to 9 inches wide can be laminated, or a series of smaller pieces can be fed through and cut apart. The machine must be preheated for a few minutes before use. This laminating machine is also used with the same film to make color-lift transparencies as explained in the section on making overhead transparencies.

Selection

Dry mounting presses are available in several models. The least expensive ones will process material up to 8 1/2 by 11 inches at one preset temperature near 225 degrees Fahrenheit. Materials up to 16 by 20 inches can be mounted in sections with some risk that lines between the sections will show. Larger and more flexible models will accommodate materials up to 26 by 32 inches (larger in sections) and with a variety of thermostatically controlled temperatures and timing devices.

Tacking irons with or without a nonsticking teflon coating are available to go with dry mounting presses. They may have adjustable temperatures also.

Dry mounting tissue is available in two versions. One produces very high quality and permanent results, but it should be used with a good press (flatiron not recommended) and no repositioning is possible. The other version contains a waxlike substance that melts at a lower temperature, and materials can be reheated and removed. The two versions cost approximately the same.

Dry mounting tissues are available in many sheet sizes up to 16 by 20 inches. They are also available in several roll sizes for large users. Large sizes can, of course, be cut into smaller sizes as needed.

The dry mounting machines with adjustable temperatures can be used for lamination with no modification other than adding a thickness of cardboard or plywood to increase the pressure.

A set of weights is very helpful for holding materials flat for a few moments after removing them from the press.

A good paper cutter that will handle the maximum size material expected is needed with all these machines. Ordinary guillotine-type machines from about 12 by 12 to 30 by 30 inches are most often used. Several other types are available that make use of a small cutting wheel or blade that travels along the paper. There is perhaps less danger with the latter if irresponsible children are involved.

Special laminating machines make the best permanent paper materials and color-lift transparencies. These machines are presently expensive, somewhat complex and made in small numbers.

Operation

A laboratory worksheet is included at the end of this section for reporting the operation of these machines.

Maintenance

The primary maintenance job with dry mounting machines is cleaning the platen that presses down on the "sandwich." If all mounting tissue is cut to exact size as described in the instructions, then no material can accumulate on the hot platen. If a clean cover sheet is put over each "sandwich," then no ink or other material can transfer to the platen. When these precautions are not taken, the platen does get dirt, ink and dry mount tissue on it. Some can be wiped off or scrubbed off with a damp cloth and soap, or even scouring powder, when the machine is cold. Thermoplastic adhesive may be removed carefully with a wooden scraper when the platen is hot. A metal scraper may easily damage the platen and be worse than leaving it dirty.

A laminating machine should be maintained by the dealer who supplied it.

Assignment XVI.   Checked _____

Name _____

Date _____

MOUNTING AND LAMINATING TEAR SHEETS

1. Sample tear sheet attached along all edges with pressure sensitive tape. Indicate tools and materials used.

   _____

   _____

   _____

2. Sample tear sheet attached with the double rubber cement method. Indicate tools and materials used.

   _____

   _____

   _____

3. Sample tear sheet dry mounted directly to this page or to a separate backing sheet. Indicate tools and materials used.

   _____

   _____

   _____

4. Laminate the lower right corner of this page or attach a sample of laminated material. Indicate the tools and materials used.

   _____

   _____

   _____

chapter 19

Room Considerations for Media Use

The traditional classroom was designed with little consideration for the regular use of modern instructional equipment and materials. Such equipment and materials as were used had to be fitted into and adapted to an environment designed primarily for lecture, recitation and reading. As new classrooms are built and old ones remodeled or rearranged, professional educators with a knowledge of media requirements must be involved from the beginning. It is now necessary to consider the requirements imposed on the structure by the media and their messages, rather than the other way around.

Figure 19.1 shows the classroom use of media as a central concern if education is going to be most interesting, efficient and permanent. Around this central concern are placed the items that need to be considered. No standard package is prescribed since the specific uses for the room determine what decisions are made about each item. The instructors who are going to use the room should determine their needs and communicate them to the building committee or architect. Too many classrooms have been designed without any instructions from those who must teach in them.

Outside light control has been a constant problem in classrooms for their entire history. Outside light varies from about 10,000 footcandles on a sunny day to near zero at other times. The source moves all over the southern half of the sky, and after a snowstorm it may appear to come intensely from everywhere. Such tremendous variations in intensity and direction obviously involve problems in control.

It was once necessary to depend on outside light for most of the classroom illumination, and large areas of glass were used, often on more than one wall and the ceiling, in order to admit enough light on dark days. All modern classrooms have an arti-ficial lighting system that will provide optimum illumination even without any outside light. It is used by most teachers all the time they are in the room regardless of outside light.

There are other arguments for outside light. Children and teachers should have constant contact with their world. It is better for their eyes. It is better for their health. It is better for their emotions. None of these arguments stands up well under scientific scrutiny.

Several classrooms have been constructed with no windows at all, and there seems to be a trend in this direction. Teachers report an immediate repulsion and a longer term liking for such rooms. The lack of distraction from outside happenings and the complete flexibility in using all the walls in many ways for many purposes seems to appeal to them. The very high costs of maintaining windows in some areas is also promoting the construction of windowless classrooms. If the classroom or presentation area is to be without windows, then it would seem wise to provide some other areas and a regular time for contact with the outside world, particularly for younger students.

Another possibility is to include one narrow window from ceiling to floor in one corner of the room. The light from this can be easily controlled with a single roller shade, and it is easily protected from vandals.

Air conditioning is a necessity if windows cannot be used for hot weather ventilation. Air conditioning is needed even with windows in many climates, and particularly if the school is used, as so many are, during the summer. Enough money can often be saved with windowless construction to pay for year-round air conditioning.

Air conditioning specifications should state the temperature, humidity and frequency of air changes that will be maintained in the room under

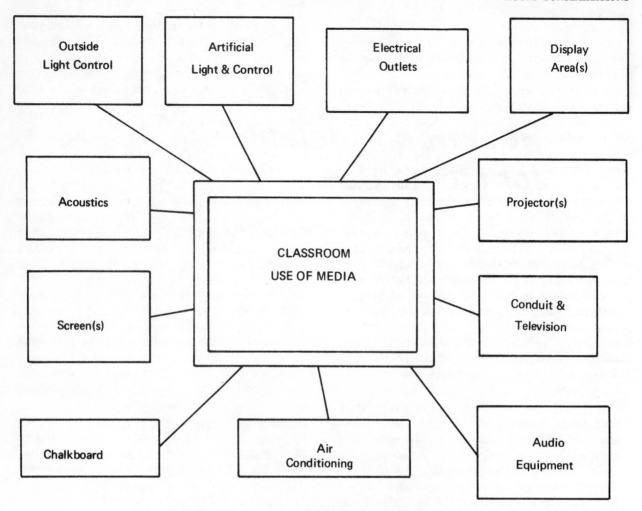

FIGURE 19.1. Classroom Considerations for Use of Media

certain outside conditions. No system guarantees optimum inside environment under the most adverse outside conditions. Some filtering system should be included in dusty locations. Another important consideration is the sound level in decibels that will be introduced into the classroom by the system. It is doubtful that any air conditioning mechanism within the room can be made quiet enough for normal classroom activities. The best systems have outside mechanisms and specially treated ductwork.

The control of light from larger windows or window walls has been solved in several ways, but none of them has been entirely satisfactory or inexpensive to install or maintain.

Drapes on tracks can be opaque to provide darkening, or translucent to control glare. The best drape installations include both of them to obtain any degree or kind of control. The drapes may be made of fabric, glass or plastic in a great variety of colors. Opaque drapes do not need to be black

or even dark in color. The drapes may be operated by direct pulling or with cords. It is essential to specify firm anchorage of tracks in unspliced or continuous lengths. Heavy-duty materials and installation are essential. If drapes are to be used, windows should not open inward unless special clearance is provided. Drapes should also be considered in relation to air inlets which may disturb them.

Venetian blinds also come in the darkening and glare control types. The darkening type is often referred to as the audiovisual blind, and it has been specially constructed so that the slats close tightly, and the four edges have been boxed in to prevent light from entering. Audiovisual blinds are usually dark in color in order to obtain the needed control. Ordinary home-type venetian blinds cannot be expected to serve long or well in a classroom.

Roller shades also come in the darkening and glare control types. Two sets may be installed as with the drapes to do all kinds of light control.

Opaque shades must be put in dark pockets or tracks all around if they are to be very effective. This complex installation is expensive and may be difficult to operate and maintain.

Light control specifications often indicate the percentage by which outside light is diminished. Ninety per cent may sound very good, but if the outside light is 10,000 footcandles with the sun shining against the shade, then 100 footcandles may result inside. A better system is to specify the footcandle level desired under normally adverse outside conditions. Levels below one footcandle are needed for good movie projection, and one-tenth footcandle is needed for opaque projection.

Control of sky lights and clerestory windows has proved particularly difficult from an audiovisual standpoint, and every effort should be made to discourage their inclusion in any new classrooms. Many existing installations have been covered with opaque material or painted over.

Inside light control does not present many problems. It is usually simply turned on or off with one or more switches. A few simple schemes will make it more flexible for use of media.

Chalkboards and display areas often need extra illumination when they are in active use. Floodlights overhead on a separate switch will easily put more than 100 footcandles on these surfaces.

Lights adjacent to the screen for the overhead projector should be on a separate switch so that only those lights are off during its use. This might be combined with the chalkboard lights previously considered.

Windowless classrooms and rooms with good darkening control need a low level lighting system in addition to the general high level system. A few incandescent lamps on a silicon-controlled rectifier (SCR) incorporated within the switch box will permit any low level illumination during projection.

Multipoint light switches at projection positions as well as the doors enable the teacher to incorporate brief sequences of projected material into a presentation or discussion.

Electrical outlets are needed all over the modern classroom as individuals and small groups use media in addition to the usual equipment stations in the front and back of the room. Twenty ampere grounded duplex outlets are generally specified. If they are 12 feet apart along each wall, then a small group and individual machine with a 6-foot cord can be used anywhere along the wall, and a large group machine with a 15-foot cord can be put almost anywhere in the room. Outlets should ordinarily be located 18 inches above the floor except

36 inches over tables and benches. A separate circuit should serve each room and be independent of the lights. The outlets are sometimes put on a switch near the door to be sure that all equipment is turned off when leaving. Outlets in the floor are generally discouraged due to difficulties in keeping dirt and water out of them.

Display areas such as bulletin boards and exhibit space within the classroom have been standard for many years, but many people now question their inclusion. One criticism involves the visual "noise" aspect in which the displayed messages may be interfering with or competing with the desired messages. Many classrooms are no longer self-contained, and several teachers find it difficult to use the same board, or even annoying to teach adjacent to competing board or exhibit. Teacher time on displays might better be put to other uses. Students might construct their displays in corridors or special display areas of the school.

Conduit means a thin-walled metal pipe for electrical wires that is buried in or fixed to the walls of the building. It is usually specified in building construction so that closed circuit television cables can be routed from a central origination or distribution point. Conduit can also be used for carrying a loudspeaker or control line from a projector in the back of the room to the front. It can be used for bringing a telephone line for a telelecture into a classroom. Classroom telephone and central sound systems usually use conduit.

The typical masonry construction of modern schools makes it very difficult to bring additional wires into a classroom already constructed. All possible needs should be anticipated and conduits installed during construction even though they may not be used for several years. Three-quarter inch inside diameter (id) conduit is most common. It may be specified to go from each designated outlet box to a central location, to a trench or false ceiling along the corridor, or it may be "looped" from box to box along a circuit. It is wise to specify that each conduit be equipped with an iron "pull wire" from box to box so that any wires can be easily pulled in at any time. The telephone company ordinarily does not share a conduit with any other service. Electrical power distribution should never share a conduit with any other service.

Chalkboard areas included in classrooms have had an upward and then downward trend. Chalkboard was once primarily used by the teacher, then expanded so that a large percentage of the class could use it for group activities, and it is now returning more to a teacher demonstration device. Many of the things that large groups of students did at the board are now done with duplicated sheets or workbooks. It is not unusual today to see large areas of chalkboard covered over with other materials. A single piece of chalkboard in the front of the classroom about 3 by 12 feet is common. The height generally varies with the height of the children, if they are going to use it. If only the teacher is to use it, the height would be similar for all grades and at about 30 inches off the floor. As overhead projectors and their large screens become standard in classrooms for the bulk of visual presentation materials, it might be wise to flank the central screen with two smaller areas of magnetic chalkboard for notices and other temporary materials related to the lesson.

Some projectors, a screen and audio equipment are now commonly considered for regular classroom storage and use rather than as temporary and portable devices to be brought in for infrequent and special use. In designing a new or remodeled classroom, consideration should certainly be given to including in the capital outlay project those machines that should be regularly used. A large and permanently mounted screen should be considered first. It would usually be attached to antikeystone brackets on the wall or ceiling and high enough so that the bottom when fully opened can be seen by all students in the room. Windowless classrooms may have one blank white wall to be used as the screen for any projector. The overhead projector is rapidly becoming a permanent classroom device ready for instant use by teacher or students. Record players, tape recorders, filmstrip projectors and cartridge motion picture machines are found in

many classrooms. All these devices are treated in detail elsewhere in this book. The essential consideration here is to determine which ones should be purchased with and made a regular part of the classroom.

The acoustical environment for teaching deserves more attention than it has usually received. It is important that the entire group can hear all messages intended without discomfort to the speakers or listeners. Outside sounds need to be attenuated by distance, absorbing materials, walls and doors or windows. No students should regularly have to sort out wanted from unwanted sounds coming into the room. Movable partitions have been particular offenders in the past. Several are now available with good decibel separation ratings. Unwanted sound should not be generated in the classroom by ventilating units, heating system, air conditioning or audiovisual machines. Sound projectors have often made so much mechanical noise that volume must be turned high to override it with consequent discomfort to several adjacent rooms. Educators have been slow to demand quiet classroom equipment. Sound in the classroom should not reverberate for too long a period. The reverberation time can be reduced by adding ceiling tile, drapes, absorbent blocks and carpeting. Carpeting also reduces the production of sound in classrooms.

Classrooms built or remodeled for modern education should be easy and comfortable places for the use of all kinds of media by the teacher and pupils. Time and energy devoted to determining how they should be constructed and equipped will be returned many fold in time, energy and discomfort saved during the years they are used.

Appendix

The actual operation and maintenance of audiovisual equipment by a group of students is assumed to be a major requirement of a course in which this book is used. A few people can do the assignments almost anywhere, but when substantial groups are concerned, a regular audiovisual laboratory is mandatory. Certain facilities, equipment and materials need to be provided for optimum operation.

Audiovisual laboratory experiences are structured in three ways. 1. A group of ten to twenty students are assigned a particular machine or operation, and they all do the same assignment at the same time with appropriate introduction, following of steps, checking of results and so forth. This scheme requires many of each type of machine, usually one for each two or three students, and enough electric power to operate the total number of machines simultaneously. Each type of machine is used only once during the course. 2. Individual study carrels are set up and equipped for each one of the sixteen laboratory assignments and one or two students at a time rotate through the entire system. Only one set of equipment and materials for each assignment is needed. Far less power is needed at any one time. Students can be self-directed and complete instructions can be included in each carrel. Single concept films, recordings, slides and the like can be programmed into the learning experience in the carrel. Minimum supervision is required. One disadvantage is that the room cannot be used for anything else. 3. A multi-purpose room is set up and equipped for regular but not necessarily exclusive audiovisual laboratory experiences. Only one of each of the sixteen equipment and materials sets is needed, and they can be stored on shelves, in drawers or cupboards, or kept in a nearby storeroom and moved to substantial laboratory tables as needed. All the heavy machines can be kept on individual projection or audiovisual stands which are then moved to any convenient location. A tote tray or drawer of materials is usually needed for each assignment. The room can be used for other purposes at other times.

The audiovisual laboratory room requires some consideration in addition to the equipment and materials needed for each assignment.

Considerably more electric power is needed than that provided for an ordinary classroom or laboratory. Many audiovisual machines require about 10 amperes, and it must be assumed that several will be operated simultaneously. It is suggested that four 20 ampere circuits be provided with only two duplex (three contact self-grounding type) outlets on each circuit to provide a total of sixteen places for power cords. If carrels are used then an outlet needs to be included in each one and preferably with a combination of low- and high-current requirements on each of the circuits. The laboratory circuits should not be shared with any adjacent classrooms. Either fuses or circuit breakers may be used and located in a nearby and unlocked access cabinet. If fuses are used, then spares are required. A master switch for all circuits in the room makes it easy to assure that all equipment is turned off at the end of the day.

The light level in the room is not critical since most images will be small and bright and observed by not more than two or three students. The opaque projector may need to be operated in a special darkened area.

Screens do not need to be provided for the projectors, other than opaque, during the experiments if there are light colored and reasonably flat wall areas available.

There is a temptation to obtain multiple use of the audiovisual laboratory equipment by scheduling it for regular classroom instruction while it is not being used for experiments. This is not recommended. The laboratory equipment is constantly tampered with in the regular course of events, and it is likely to prove unreliable and exasperating to teachers who try to use it. It is likewise exasperating to the laboratory instructor and students when equipment must be found before it can be used.

A sink in the room is helpful, but not essential unless regular color-lift transparencies or other fluid process materials are to be made.

Security provisions may be needed since the collection of equipment and materials is valuable, and several items have direct application in home entertainment, hobby workshops and so forth. There is also a tendency for equipment and materials intended for laboratory instruction to be appropriated for regular classroom instruction.

Each assignment requires a collection of advertising material from a variety of companies so that the students can study the various brands, models and accessories available in addition to the few actually available for operation.

These sheets can be obtained from local or regional audiovisual dealers, or they may be obtained by writing to the manufacturers at the address given in the *Audio-Visual Equipment Directory* published each year by the National Audio-Visual Association, Inc., 3150 Spring Street, Fairfax, Virginia, 22030. Several copies of this directory should also be in the laboratory so that most available machines can be seen. Each machine should also be supplied with the specific instruction sheet that came with it. It might be wise to mount or laminate the sheet or sheets. Additional sheets can and should be obtained from the manufacturer and kept in a special file for use when the regular ones are lost or worn out. Service or maintenance booklets can be obtained for some machines if additional information is desired.

Some equipment and materials should be available in a convenient location for several of the laboratory assignments. The following items could be kept in a special drawer or tote box and labeled "maintenance materials":

1. soft cloth for cleaning
2. soft brush for cleaning (paint or poster type)
3. lens tissue
4. electric motor or household oil
5. screwdrivers (several sizes of regular and Phillips head)
6. pliers (5" or 6")
7. diagonal cutters (5" or 6")
8. soldering iron, solder and flux or paste

9. extension cords (heavy duty)
10. yardsticks and/or carpenter's 6-foot rules (several needed)

The following equipment and materials are keyed to the sixteen laboratory assignments and might be considered as a minimum listing. Additional or specialized items should be added for special purposes or situations.

I. Basic Optical System
   A. Equipment
      1. Clear glass (not frosted) 10-40 watt 115-120 volt incandescent lamp in fixture or socket
      2. Round reading or magnifying glass
      3. Complete filmstrip or slide projector (can be an inexpensive, old or low power model)
      4. Small tripod screen with any type surface
   B. Materials
      1. Filmstrip or piece of filmstrip
      2. 2 by 2 slides
      3. 3 by 4 slides
      4. Advertising literature on representative filmstrip, slide, overhead and opaque projectors

II. Standard Lantern Slide Projectors
   A. Equipment
      1. Standard lantern slide projector with slide carrier
      2. Adaptor for 2 by 2 or other smaller slides
   B. Materials
      1. Instruction sheets for projector used
      2. Several photographic and handmade lantern slides
      3. Advertising literature for various lantern slide projectors

III. Projection Screens
   A. Equipment
      1. Mat surface tripod screen with antikeystone device
      2. Beaded surface tripod screen with antikeystone device
      3. Lenticular, sunscreen or other special screen surface
      4. Projector to provide screen image (may be borrowed temporarily from another laboratory station)
   B. Materials
      1. Instruction sheets for screens used
      2. Slide or film for projector used
      3. Samples of various screen fabrics and surfaces for front projection

4. Sample of translucent material for rear projection
5. Advertising literature on screens

IV. Overhead Projectors
  A. Equipment
    1. Overhead projector or several makes and models to illustrate various optical systems in common use
    2. Antikeystone screen
  B. Materials
    1. Instruction sheets for machine(s) used
    2. Prepared overhead projection transparencies
    3. Plastic roll and/or blank plastic sheets
    4. Pens and pencils for marking on plastic
    5. Water, lighter fluid and tissue for cleaning transparencies
    6. Advertising literature to show various makes and models of overhead projectors
    7. Advertising literature to show materials for local production of transparencies
    8. Advertising literature to show commercially prepared transparencies

V. Overhead Transparency Makers
  A. Equipment
    1. Thermal copying machine
    2. Diazo copying machine
    3. Primer typewriter
    4. Various machines for making transparencies by the electrostatic, color-lift and photoreflex systems (optional)
  B. Materials
    1. Instruction sheets to go with machines
    2. Plastic sheets, markers, tapes, rub-on letters, etc., for local production of transparencies
    3. Masters and tear sheets for thermal copying
    4. Masters for diazo copying
    5. Thermal copy film
    6. Diazo copy film
    7. Tracing paper, markers, tapes, etc., for local production of masters
    8. Cleaning materials or kits recommended or provided by makers of particular machines used
    9. Advertising literature to show various makes and models of transparency makers

VI. Opaque Projectors
  A. Equipment
    1. Opaque projector
    2. Tripod screen
  B. Materials
    1. Instruction sheets for projector used
    2. Various tear sheets for projection
    3. Small flat objects for projection
    4. Book (of no value) for projection
    5. Advertising literature to show various makes and models

VII. Filmstrip and 2 by 2 Slide Projectors
  A. Equipment
    1. Combination filmstrip, 2 by 2 slide projector or separate filmstrip and manual slide projectors
    2. Automatic and/or remote control filmstrip and/or slide projectors (optional)
  B. Materials
    1. Instruction sheets for machines used
    2. Sample filmstrips
    3. Sample 2 by 2 slides (various formats and mounts)
    4. Sample filmstrip cartridges and slide trays
    5. Advertising literature to show various makes and models

VIII. 2 by 2 Slidemakers
  A. Equipment
    1. Camera and flash device for making 2 by 2 slides
    2. Close-up stand and supplementary lenses for camera
    3. Lights or flash arrangement for close-up stand
    4. Light meter (may be included in camera)
  B. Materials
    1. Instruction sheets for equipment used
    2. Film for camera used
    3. Flashbulbs or cubes
    4. Tear sheets to photograph
    5. Advertising literature to show various makes and models of cameras and close-up stands

IX. Public Address System
  A. Equipment
    1. Public address system or separate microphone, amplifier and loudspeaker
    2. Tuner, record player, earphones, various microphones and various loudspeakers (optional)

B. Materials
  1. Instruction sheets for equipment used
  2. Record (if record player is to be used)
  3. Advertising literature to show various makes and models of microphones, amplifiers, loudspeakers, complete systems and accessories.

X. Record-Transcription Players
  A. Equipment
    1. Record-transcription player (a versatile and high quality model is needed rather than an ordinary record player)
    2. Stylus lens and pressure gage (optional)
    3. Microphone, extension loudspeaker, earphones (optional)
  B. Materials
    1. Instruction sheets for machine used
    2. Variety of discs including several diameters and speeds and both groove sizes
    3. Advertising literature to show various makes, models and accessories

XI. Tape Recorders
  A. Equipment
    1. Tape recorder (standard classroom model)
    2. Professional tape recorder (optional)
    3. Battery-operated portable tape recorder (optional)
    4. Patch cord(s) for recording from record player, tuner or another recorder (optional)
  B. Materials
    1. Instruction sheets for machines used
    2. Tape of various thicknesses on various size reels
    3. Tape in a cassette or cartridge
    4. Tape recorder cleaning kit
    5. Advertising literature to show various makes, models and accessories

XII. Motion Picture Projectors
  A. Equipment
    1. Manual threading 16mm sound projector(s) (several makes and models are highly desirable)
    2. Self-threading 16mm sound projector(s) (optional)
    3. Eight-millimeter projector(s) (optional)
  B. Materials
    1. Instruction sheets for machines used

    2. Sixteen-millimeter film
    3. Eight-millimeter film in various formats
    4. Aperture cleaning brush
    5. Spare exciter lamp
    6. Take-up reels of various sizes
    7. Advertising literature to show various makes, models and accessories

XIII. Single Concept Moviemakers
  A. Equipment
    1. Automatic super-8 movie camera
    2. Camera light matched to camera for indoor filming
    3. Titling kit to prepare titles (optional)
    4. Projector for super-8 film
  B. Materials
    1. Instruction sheets for equipment used
    2. Unexposed film in cartridge for camera to be used
    3. Exposed film on reel or in cartridge (depending on projector used)
    4. Cartridge loading kit (depending on system used)
    5. Advertising literature on various super-8 cameras, projectors and accessories

XIV. Film, Tape and Electrical Cord Repairs
  A. Equipment
    1. Sixteen-millimeter film splicer
    2. Single-edge razor blades
    3. Scissors
    4. Splicing blocks for 16mm and 8mm film and sound tape
    5. Audio tape splicer (Gibson Girl)
    6. Tools for cord repairs (pliers, knife, screwdriver, soldering iron, also included in maintenance drawer)
    7. Replacement 120V AC cord connector
  B. Materials
    1. Instruction sheets for splicers used
    2. Scrap film for splicing
    3. Film splicing cement
    4. Mylar patches for 16mm and 8mm film repair
    5. Pressure sensitive transparent tape 3/4 inches wide
    6. Audio splicing tape
    7. Scrap wire for splicing and connector attachment
    8. Solder, solder flux or paste, electrical tape
    9. Advertising literature on various makes and models of splicers

XV. Closed Circuit Television Systems
  A. Equipment
    1. Complete and integrated basic closed circuit television system
    2. Supplementary lenses, lighting, microphones, mixers, special effects generator, tripod, monitors, etc. (optional)
  B. Materials
    1. Instruction booklets for equipment used
    2. Recording tape to fit recorder used
    3. Recorder cleaning kit for specific recorder used
    4. Advertising literature on various closed circuit television systems and accessories
XVI. Mounting and Laminating Machines
  A. Equipment
    1. Paper cutter (at least 12″ x 12″)
    2. X-acto knife with #11 blade
    3. Dry-mounting press
    4. Flatiron (discarded household type)
    5. Laminating machine (optional)
  B. Materials
    1. Instruction sheets for press and laminator
    2. Construction or other heavy mounting paper
    3. Rubber cement and thinner
    4. Wax paper
    5. Pressure sensitive mounting tapes
    6. Dry-mounting tissue
    7. Laminating films
    8. Advertising literature on various mounting and laminating devices and materials

REFERENCE MATERIALS

A. General Audiovisual References
BROWN, J. W., R. B. LEWIS and F. F. HARCLEROAD, *A-V Instruction: Media and Methods*, 3rd ed. New York: McGraw-Hill Book Company, 1968.
DALE, EDGAR, *Audio-Visual Methods In Teaching* rev. ed. New York: Dryden Press, 1954.
ERICKSON, C. W. H., *Fundamentals of Teaching with Audiovisual Technology*, New York: The Macmillan Company, 1965.
KINDER, J. S., *Audio-Visual Materials and Techniques* 2nd ed., New York: American Book Company, 1957.
THOMAS, R. MURRAY and SHERWIN G. SWARTOUT, *Integrated Teaching Materials* rev. ed., New York: David McKay Co., Inc., 1963.
WITTICH, WALTER A. and CHARLES F. SCHULLER, *Audio-Visual Materials: Their Nature and Use* 4th ed., New York: Harper & Row, Publishers, 1967.

B. Local Production
KEMP, JERROLD E. et al., *Planning and Producing Audio-Visual Materials*, 2nd ed. San Francisco: Chandler Publishing Co., 1968.
MINOR, ED., *Simplified Techniques for Preparing Visual Instructional Materials*, New York: McGraw-Hill Book Company, 1962.
MONLAN, JOHN E. *Preparation of Inexpensive Teaching Materials*, San Francisco: Chandler Publishing Co., 1963.
C. Equipment
Audio-Visual Equipment Directory, Annual Editions, National Audio-Visual Association, Fairfax, Virginia.
DAVIDSON, RAYMOND L., *Audiovisual Machines*, Lubbock, Texas: The Texas Tech Press, 1964.
EBOCH, SIDNEY C., *Operating Audio-Visual Equipment*, 2nd ed. San Francisco: Chandler Publishing Co., 1968.
PULA, FRED J., *Application and Operation of Audiovisual Equipment in Education*, New York: John Wiley & Sons, Inc., 1968.
D. School Design for Media
DE BERNARDIS, AMO *et al.*, *Planning Schools for New Media*, Portland, Oregon: Portland State College, 1961.
Design for ETV: Planning for Schools with Television, Educational Facilities Laboratories, New York, 1960.
Educational Facilities with New Media, Department of Audiovisual Instruction, National Education Association, Washington, D. C., 1966.
FARIS, GENE, *et al.*, *Improving the Learning Environment*, USOE, OE-34031, 1963.
E. Audiovisual Journals
Audiovisual Instruction, DAVI-NEA, 1201 16th St., NW, Washington, D. C. 20036.
AV Communication Review, DAVI, 1201 16th St., NW, Washington, D. C. 20036.
Education Age, Visual Products Div., 3M Co., 3M Ctr., St. Paul, Minn. 55101.
Educational Screen & Audiovisual Guide, 434 S. Wabash, Chicago, Ill. 60605.
Educational Technology, 198 Ellison St., Paterson, N. J.
Film News, 250 West 57th St., N.Y., N.Y. 10019.
Media and Methods, 134 N. 13th St., Philadelphia, Pa. 19107.
Training in Business & Industry, 33 West 60 St., N.Y., N.Y. 10023.
F. Postal Rates for Audiovisual Materials
Special postal rates for library and audiovisual materials have been in effect for many years. These provide a governmental subsidy for educational materials so that they will be more available to users. The rates and rules change from time to time, so those who mail materials should check regularly with the local postmaster. There is generally a special rate for educational materials shipped by anyone to anyone, and a still lower rate for such materials sent to or from an educational institution. In order to qualify for the special rates, special labels or rubber stamps should be prepared with a designation such as: LIBRARY MATERIALS—

THIS PACKAGE MAY BE OPENED FOR POSTAL INSPECTION or whatever is suggested by the local postmaster.

G. Preparing Specifications for Audiovisual Equipment Purchase

When educators plan to purchase audiovisual equipment, they need to know how well a certain make and model at a certain price will fit their particular needs in a particular setting. The following questions should be considered:

1. How safe is it?
2. How bright is the image? Optional lamps? Optional lenses?
3. How faithful is the image?
4. What constraints are made on image size and distance? Options?
5. How much light is spilled and where?
6. What volume of sound is produced?
7. What fidelity of sound is produced?
8. What extraneous sound is produced?
9. How hot does it get? Where? When?
10. What is its size and weight?
11. What is its appearance?
12. What convenience features are standard and optional?
13. What are its power requirements?
14. How compatible is it with other machines? Connectors? Lamps? Lens Diameter? Cartridges? Etc.
15. What guarantee is provided and by whom?
16. What service facilities are available and where?
17. Does it damage or wear the materials used with it?
18. What service is needed at what intervals?

Index